Leibniz: A Very Short Introduction

VERY SHORT INTRODUCTIONS are for anyone wanting a stimulating and accessible way into a new subject. They are written by experts, and have been translated into more than 40 different languages.

The series began in 1995, and now covers a wide variety of topics in every discipline. The VSI library now contains over 450 volumes—a Very Short Introduction to everything from Psychology and Philosophy of Science to American History and Relativity—and continues to grow in every subject area.

Very Short Introductions available now:

Available soon:

For more information visit our website

www.oup.com/vsi/

Maria Rosa Antognazza

LEIBNIZ

A Very Short Introduction

OXFORD
UNIVERSITY PRESS

OXFORD
UNIVERSITY PRESS

Great Clarendon Street, Oxford, OX2 6DP,
United Kingdom

Oxford University Press is a department of the University of Oxford.
It furthers the University's objective of excellence in research, scholarship,
and education by publishing worldwide. Oxford is a registered trade mark of
Oxford University Press in the UK and in certain other countries

Published in the United States of America by Oxford University Press
198 Madison Avenue, New York, NY 10016, United States of America

British Library Cataloguing in Publication Data
Data available

Library of Congress Control Number: 2016939030

ISBN 978-0-19-871864-2

Printed and bound by
CPI Group (UK) Ltd, Croydon, CR0 4YY

To Howard

Contents

Leibniz

Preface

A very short introduction to Leibniz is a very large challenge. For one thing, there are no easy or obvious entry points into Leibniz's thought. Moreover, since all aspects of his thought are mutually dependent, every starting point seems to require the knowledge of some other, or an understanding of his broader philosophical or not strictly philosophical commitments. To make matters worse, Leibniz himself left no single, comprehensive, and systematic exposition of his philosophical thought. Without it, generations of Leibniz scholars have had to tease out an overview of his gradual and complex intellectual evolution largely from his sprawling correspondence and innumerable working papers.

As a result of these difficulties, specialists continue to puzzle over Leibniz's intellectual legacy 300 years after his death. As previously unpublished letters and other writings have emerged from the seemingly unfathomable depths of his archive, his persona has been shaped and reshaped by successive waves of interpretation. Today, debates on both his mature philosophy and his philosophical development are more vigorous than ever. Was he an idealist or a realist about bodies? Did he even reach a final view on the matter, or did he continue to experiment intellectually until his dying day? More generally, were his mature philosophical views broadly consistent? Or did he hold different and inconsistent philosophical models? If the latter, were these different models held at the same time, or at different stages of his development? Did he succeed in defending contingency, or did he slide inexorably into necessitarianism? Is his claim that this is the best of all possible worlds a plausible basis for a defence of God's justice and goodness in the face of so much evil in the world? Or should it be discarded as facile optimism? These and many other interpretative questions are still fiercely debated among specialists. Small wonder, then, that novices may fear to venture into the labyrinth of Leibniz's thought, where the danger of becoming lost seems to lurk around every corner.

At the heart of the problem is the fact that many of Leibniz's philosophical views are so counter-intuitive that the unwary reader may be tempted to dismiss them as a philosophical fairy-tale. He maintained, for instance, that there cannot be two identical grains of salt or blades of grass in any of the infinitely many worlds which are possible. In later years, to give another example, he claimed that everything results from mind or mind-like simple substances or 'monads'. Yet Leibniz was anything but an undisciplined fantasist. He was a mathematician and logician of the first magnitude, entirely committed to scientific enquiry. Indeed, it is perhaps the very technical rigour and sophistication of his philosophy that poses the ultimate challenge to the uninitiated reader. For his own part, Leibniz was fully convinced that his metaphysics (that is, roughly, his theory of what is ultimately real) was the most plausible explanation of what we experience in the world; and some of his most counter-intuitive conclusions have been vindicated by the strange world of modern physics, so different from the more immediately intelligible (but wholly inadequate) mechanical natural philosophy of Descartes.

It hardly needs to be said that no *very short introduction* can do full justice to the range of Leibniz's philosophical thought and its evolution over time, or to the merits of different interpretations of that thought and development. Still more impossible is to complement an adequate outline of his philosophy with an equal treatment of his wide-ranging contributions to an exceptional variety of other intellectual fields, to say nothing of his many practical, legal, and diplomatic activities. Rather than offering a bald summary of everything he attempted or accomplished, this slim volume aims to serve the core community of interest on Leibniz, placing an introductory outline of some central features of his philosophy in the context of his overarching intellectual vision and aspirations. Trying to understand who Leibniz was, what his fundamental commitments were, and what he wanted to achieve (Chapters 1–3), will be our way to begin a very short journey through possible worlds and what is ultimately real in the best of them all (Chapters 4–9).

Before embarking on this journey, I would like to thank Niccolò Guicciardini, Paul Lodge, Massimo Mugnai, and the anonymous reviewers secured by Oxford University Press for their very helpful feedback on my initial outline and the complete manuscript. Thanks are also due to Jenny Nugee and Andrea Keegan of OUP for their expert advice throughout the process which has seen this project through the press. As always, my greatest debt is to Howard Hotson for his unfailing support and perceptive reading of the manuscript. Any shortcomings or unbalances in my account of Leibniz remain of course my sole responsibility.

List of illustrations

1. Portrait of Leibniz by Andreas Scheits (1703).

Chapter 1
Who was Leibniz?

Gottfried Wilhelm Leibniz (1646–1716) lived an extraordinarily rich and varied intellectual life in troubled times. Although remembered as a great thinker, he was a man who more than anything else wanted to *do* certain things, namely to improve the life of his fellow human beings through the advancement of science, and to establish a stable and just political order in which the divisions amongst the Christian churches could be reconciled. A surprising number of his apparently miscellaneous endeavours were aspects of a single master project tenaciously pursued throughout his life: the systematic reform and development of all the sciences, to be undertaken as a collaborative enterprise supported by an enlightened ruler. These theoretical pursuits were in turn ultimately grounded in a practical goal: the improvement of the human condition in a well-ordered world, and the celebration thereby of the glory of God in His creation. 'To contribute to the public good and to the glory of God', Leibniz maintained, 'is the same thing' (A I, 18, 377); and this aim he intended to pursue relentlessly 'by means of useful works and beautiful discoveries' (A I, 2, 111).

The scintillating intellectual development of a young man

Born in Leipzig (Germany) on 1 July 1646 to a family of moderately comfortable academics and jurists, Leibniz was raised in a stronghold of Lutheranism just recovering from a protracted war which did not officially end until two years after his birth. The Peace of Westphalia, signed at the end of the Thirty Years War (1618–48), allowed the Holy Roman Empire to regroup as a somewhat precarious political entity formed of hundreds of imperial estates with varying claims to autonomous government and religious freedom. One key achievement at Westphalia was to hammer out rules for the permanent and peaceful coexistence of the three main Christian denominations: Lutheran, Reformed, and Roman Catholic. Leipzig was one of the two main cities of Saxony, a bastion of Lutheranism which benefited from the success of its duke in allying with the Roman Catholic Habsburgs (the ruling imperial family) against the even more hated Calvinists.

Leibniz's father, a professor of moral philosophy at Leipzig University, died in 1652, leaving the young Gottfried Wilhelm and his sister in the care of their mother, and in a house full of books. The presence of so many books and the absence of their owner apparently gave the young boy unusual opportunities to read widely without close paternal supervision. The library included not only the staples of an academic family of strict Lutherans but also more varied deposits left by the Leipzig publisher and bookseller whose daughter had been married to Leibniz's father until her death in 1643. By reading through parallel literature on contested issues, the young Leibniz developed a conciliatory and ecumenical outlook strikingly at odds with the narrowly confessional views of his immediate and extended family. This explains why Leibniz, despite attending the University of Leipzig and its distinguished preparatory school, the Nikolaischule, always regarded himself as an 'autodidact'. Yet his youthful outlook was shaped also by some inspirational university professors in Leipzig and (for the summer semester of 1663) in Jena, who passed on to him a complex mixture of Scholastic Aristotelianism and heavily theologized Platonism, as well as a fascination with mathematics.

After gaining Bachelor and Master degrees in Philosophy and a Bachelor degree in Law, and writing one of his first brilliant works, the *Dissertation on Combinatorial Art* (1666), Leibniz enrolled in October 1666 at the University of Altdorf in order to expedite his doctorate in Law. Just over a month later, he passed the examination of his thesis with great distinction—so great, in fact, that he was immediately offered a position. But the 20-year-old had much grander plans than settling as a law professor in a provincial university. He had already decided that his life's calling was to help piece back together the fragments of a world, torn apart by religious, political, and intellectual divisions, into a new, universal synthesis for the glory of God and the happiness of humankind.

The first step on this steep road, the young Leibniz decided, was to broaden his horizons by undertaking a European grand tour via the Rhine and Holland. In the event, this led no further than Frankfurt and the nearby Catholic archiepiscopal seat in Mainz. Yet, thanks to the complicated confessional geography of the Empire, this short journey broadened his horizons more than he might have hoped. At the tolerant and intellectually lively court of the archbishop, Leibniz met patrons who opened for him new perspectives. In Mainz, he conceived the first version of the overarching plan of reform and advancement of all the sciences which was to traverse his entire life as a sort of Ariadne's thread. The 'Catholic' (that is, universal) 'Demonstrations' envisaged by this youthful plan covered natural theology (the 'Demonstration of God's Existence' and the 'Demonstration of the Immortality and Incorporeity of the Soul'), revealed theology (the 'Demonstration of the Possibility of the Mysteries of the Christian Faith'), and the 'Demonstration of the Authority of the Catholic [that is, universal] Church' and 'of the Authority of Scripture'. Most importantly, as *prolegomena* to the catholic demonstrations proper, the young Leibniz listed the development of the 'elements of philosophy', namely the first principles of metaphysics (*de Ente*), of logic (*de Mente*), of mathematics (*de Spatio*), of physics (*de Corpore*), and of

3

ethics and politics or 'practical philosophy' (*de Civitate*) (A VI, 1, 494). In brief, this plan provided a first outline of a systematic encyclopaedia of the sciences conceived in support of stable features of Leibniz's thought such as the existence of God and the immortality of the soul.

His patron, Baron Johann Christian von Boineburg, was also instrumental in facilitating a four-year trip to Paris, interspersed by two short visits to London, which proved transformative. Leibniz left for the French capital in March 1672 and returned to Germany in December 1676 as an accomplished mathematician who had invented nothing less than the infinitesimal calculus (October 1675), been elected Fellow of the Royal Society (April 1673), mixed with leading scientists in Paris and London (including Christiaan Huygens and Robert Boyle), and discussed philosophical matters with Baruch Spinoza on his way back via Holland (November 1676).

Stretching boundaries and thinking large for the common good

After this youthful period of brilliant intellectual developments and encounters, the next forty years of Leibniz's life were characterized by his attempts to stretch the narrow brief of his official duties to further his all-embracing plan of reform. In mid-December 1676, unable to prolong any longer an already circuitous journey back to his native country, Leibniz arrived in the provincial German city of Hanover to serve as court counsellor and librarian to Duke Johann Friedrich, the head of a cadet line of the Guelf family (see Figure 2). The Hanoverian court was a far cry from the scientific communities of Paris and London, but the duke was a congenial patron with whom Leibniz could share his grand vision of the 'catholic demonstrations'. Soon after he presented his plans to the duke, however, Johann Friedrich died, on 28 December 1679.

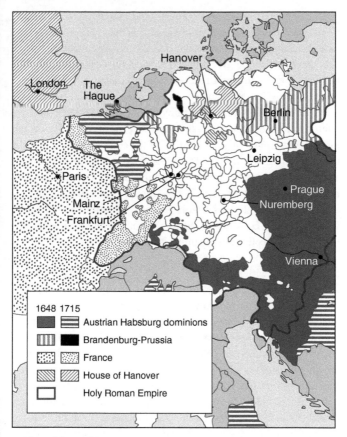

2. **Map of Central Europe.**

The two successive dukes of Hanover, Ernst August (1629–98)
and his son Georg Ludwig (1660–1727, from 1714 also George I of
Great Britain), had little time for Leibniz's enlightened appeals for
the common good. Rather than advancing science and knowledge,
their aim was to advance the standing of their territory from
junior duchy to major player in the imperial hierarchy. For such a

purpose, the legal learning, ingenuity, and industry of their clever if idiosyncratic court counsellor were nevertheless invaluable. Leibniz also promised to solve stubborn problems affecting the productivity of their silver mines (the most important source of income for the Hanoverian dukes) through daring technical innovations. When it became clear that the theoretical elegance of his engineering was not matched by practical results, Leibniz offered to further their interests by mining the archives of Europe instead, in search of historical proof of the connection between the Guelf and the Este families. This quest was no idle genealogical curiosity. The dynastic connection established by Leibniz's research proved a key plank in the Guelf dukes' claim to an elevated status of electors, that is, as one of the nine senior princes who elected the Holy Roman Emperor himself and formed the most powerful college in the imperial Diet or Reichstag.

Never content with half measures, Leibniz extended his historical task into a fully-fledged history of the Guelfs beginning, quite literally, from the dawn of time with the prehistory of the earth and the migrations of people. But historical research also provided him with the excuse to leave Hanover for extensive travels abroad. For two and a half years (November 1687–June 1690), deaf to the enquiries of his employers as to his precise whereabouts, he toured southern Germany, Austria, and Italy, seizing every opportunity to meet European savants, scientists, and technicians. Once back in Hanover, his habit of disappearing from his desk for months or even years at a time continued, undeterred by the displeasure of Ernst August and, especially, Georg Ludwig. During his long unauthorized visit to the imperial court in Vienna (December 1712–September 1714), his employer finally resorted to freezing Leibniz's stipend, and expressed his exasperation in a letter of warning to the empress herself:

> Leibniz, by his native disposition, wants to achieve everything; and he therefore delights in endless correspondence and shuttling to and fro, striving to satisfy his insatiable curiosity; but he has neither

talent or inclination to pull anything together and bring it to a close. It would be regrettable if the Elector were to lose him [as a counsellor] and yet the Emperor to gain nothing useful from him. (Doebner, 217)

However unsympathetic, this portrait is not unjust in highlighting Leibniz's inclination to conceive grand new schemes rather than finishing older ones. Yet even the dour Georg Ludwig could see that his uniquely gifted adviser was irreplaceable as he was ungovernable. Unlike many contemporary savants, Leibniz was neither a gentleman of leisure, with independent means for supporting his scientific investigations, nor a clergyman or a university professor, whose day job included writing and teaching theology, philosophy, mathematics, and the like. The theoretical writings which claim the attention of modern scholars were naturally perceived by his main Hanoverian employers as an irritating distraction from his duties as legal adviser, librarian, historiographer, and, at a stretch, mining engineer.

For his part, Leibniz understood himself as a sort of science and culture minister without portfolio in perennial search of an enlightened ruler who would provide the institutional backing necessary for major enterprises such as the establishment of academies of sciences, the reform and rationalization of public administration, the establishment of medical statistics geared at the prevention of illnesses, a public health system, life insurance, or a system of financial protection for widows and orphans. While the Hanoverian dukes expected, from their doctor in Law, painstaking legal paperwork to back up their narrow dynastic interests, Leibniz was keen to give them wide-ranging political advice on matters closer to his own heart, above all, how to secure stability in the Holy Roman Empire, and hence a peaceful political order. For all his yearning for Paris and London, Leibniz was deeply committed to the Empire and to its role in Europe. Unlike his contemporary countryman, Samuel Pufendorf (1632–94), who described the Empire as a multi-headed monster, Leibniz regarded it (rather like the latest historical scholarship today) as an alternative model of shared sovereignty in which territorial autonomy could be combined with a central imperial authority, and the main Christian confessions could cohabit peacefully in a balanced, representative Reichstag. A strong Holy Roman Empire was in his view a key condition for a lasting peace in Europe, opposing the expansionism of Louis XIV's France on the west flank and the advances of the Ottoman Empire to the east. His indefatigable work to secure a Protestant succession to the English crown, and hence the accession to the throne of the (initially reluctant) Hanoverians, was in keeping with his commitment to a balance of power in Europe amongst the Christian confessions. Short of achieving the church reunification for which he passionately campaigned throughout his life, a Protestant succession in Great Britain was the best means of counter-balancing the recent political gains of the Catholic side.

All things considered, although he never tired of searching for other patrons for his plans and proposals, the house of Hanover provided him with the security of a (fairly regularly) paid position, court benefits, and a platform for his political vision and scientific ventures. In Hanover he was fortunate to count on the protection and intellectual sympathy of two extraordinary women, Sophie von der Pfalz (wife of Ernst August and mother of Georg Ludwig) and Caroline von Ansbach (daughter-in-law of Georg Ludwig, and future princess of Wales). In the nearby Guelf Duchy of Wolfenbüttel, he also enjoyed the esteem and friendship of the local dukes, Anton Ulrich and Rudolph August, for whom he directed the impressive ducal library. In Berlin, at the increasingly powerful royal court of Brandenburg-Prussia, he had his most cherished friend and unwavering ally in Sophie Charlotte, daughter of Sophie and wife to the Prussian king.

It was in Berlin that he eventually succeeded in founding an academy of sciences in 1700, which he later tried to replicate in Saxony and in Vienna. The court of the Habsburg emperors in Vienna was in fact one of his recurrent targets for patronage, where he was welcomed by influential supporters such as Prince Eugene of Savoy. The list of promising contacts from whom he hoped to receive backing for his overarching proposals extended in the last phase of his life to the Russian tsar, Peter the Great. Support for his reforms and scientific ambitions never materialized on the scale he had wished, but his shuttling between Hanover, Wolfenbüttel, Berlin, and Vienna brought him the intellectual stimulation he craved, complementing the exceptionally broad network connecting him over the years to around 1,300 correspondents.

Leibniz was gregarious by nature and a great conversationalist, who treasured the constant distractions of lively intellectual exchange. 'So one gradually advances,' he wrote to one of his correspondents in 1694, 'responding to the demands of the moment' (A I, 10, 143). Indeed, one of the most distinctive features of his intellectual production was the scope and depth of his correspondence. Some of the most illuminating accounts of his views are to be found in discussions with epistolary partners who forced him to refine aspects of his theories, or demanded that they be restated in clearer terms. He debated metaphysical, physical, and mathematical issues with Johann Bernoulli (1667–1748), Burchard de Volder (1643–1709), Bartholomew Des Bosses (1668–1738), Christian Wolff (1679–1754), and many others. His correspondence with the Newtonian Samuel Clarke (1675–1729), mediated by Princess Caroline, constituted a treasure trove of mature views on a wide range of issues (including, for instance, the nature of space and time), even while he was still smarting from the wounds inflicted by the protracted priority war on the discovery of the calculus which pitted Leibniz against Isaac Newton (1642–1727) and his army of followers.

It was also in conversation with others that some of his best-known pieces developed. In 1686 he penned a *Discourse on Metaphysics* which furnished issues for discussion over the following few years in a rich (if sometimes frustrating) correspondence with one of the leading intellectuals of the age, the French theologian and philosopher, Antoine Arnauld (1612–94). During his extended visit to Italy in 1689–90, stimulated by the meetings of the Accademia Fisico-Matematica in Rome, he wrote his *Dynamics*, placing the notion of force at the heart of physics. Between 1703 and 1705, he composed the *New Essays*, a virtual conversation with John Locke (1632–1704) about his *Essay Concerning Human Understanding* (1689). A lifetime of debate and reading, supplemented by more recent conversations with Sophie Charlotte and her entourage in Berlin, came together in the *Theodicy* of 1710. Requests for clarifications of his metaphysical theories from one of his correspondents, Nicolas Rémond, and from Prince Eugene of Savoy prompted Leibniz to compose in 1714 two agile pamphlets, the *Principles of Nature and of Grace* and the *Monadology*.

Another distinctive feature of his intellectual production is the relative paucity of writings printed in his lifetime. To be sure, Leibniz published plenty of landmark mathematical and scientific papers in the learned journals of the time, including the *Nova methodus pro maximis et minimis* (1684), that is, his first public presentation of the calculus; the *Brevis demonstratio erroris memorabilis Cartesii* (1686), contending, against René Descartes (1596–1650), that force, not motion, is conserved; the *Tentamen de motuum caelestium causis* (1689), proposing Leibniz's own cosmological system; and the *Specimen dynamicum* (1695), introducing a new branch of physics devoted to the study of forces which Leibniz christened 'dynamics'.

In addition to the *Theodicy*, a handful of philosophical essays disclosing key aspects of his epistemology and metaphysics were also published in his lifetime, notably, the *Meditations on Knowledge, Truth, and Ideas* (1684), *On the Emendation of First Philosophy, and on the Notion of Substance* (1694), the *New System of the Nature of Substances* (1695), and *On Nature Itself*

(1698). But a vastly greater amount of notes, sketches, essays, and even book-length discussions remained piled in manuscript form in his Hanoverian quarters, including—astonishingly—such key writings as the *Discourse on Metaphysics*, the *New Essays*, the *Principles of Nature and of Grace*, and the *Monadology*. The modern critical edition of Leibniz's philosophical texts written between 1677 and June 1690, for instance, contains 522 pieces totalling over 3,000 pages. Only three of these texts, adding up to scarcely twenty pages, were published by Leibniz himself.

Small wonder, then, that Leibniz once wrote that those who knew him through his published writings hardly knew him at all. The gradual expansion of the main collections of Leibniz's writings which have appeared at regular intervals during the three centuries since his death helps explain why the outline of Leibniz's thought has so often changed shape in the mirror of modern scholarship. From his publications, Leibniz appeared to the learned readership of his day primarily as a mathematician and a scientist. The full depth of his philosophical insights and the astoundingly broad range of his intellectual interests were glimpsed only by his most regular correspondents and trusted friends, who spread rumours of his unique gifts through the contemporary republic of letters.

There was unquestionably an intensely private as well as an engagingly public side to his intellectual life; but these were two sides of the same coin, not the Janus faces ascribed to Leibniz by those, like Bertrand Russell, who have argued that he held a sincere philosophy in private and a false one in public, adopted to please the court. Nor should we forget that the Hanoverian court counsellor was largely free from the compulsion to publish, so inescapable in academic life today. It was not by placing a metaphysical paper in a top international journal that Leibniz would have pleased his employers. Instead, the value of each publication was measured with reference to his main aim and objectives. Leibniz was a man of synthesis and reconciliation. His overarching goal was the improvement of the human condition.

To his mind, theoretical reflections on logic, mathematics, metaphysics, physics, ethics, and theology were ultimately in the service of life and aimed at the happiness of humankind. If he perceived that some of his most striking philosophical views would have bred disagreement and misunderstanding rather than contributing to his main goal of reconciling opposing parties—whether theologians belonging to different confessions, philosophers with clashing doctrines, or politicians able to support his projects—he preferred to keep them to himself. Because his writings were ultimately part of an all-embracing philosophical, theological, and political project, and even the most abstract work of metaphysics could have implications for the fragile confessional situation in Germany, Leibniz was often reluctant to release them prematurely while still sketchy or potentially controversial.

In August 1714, following the death of Queen Anne, the British crown passed to Georg Ludwig of Hanover. Upon receiving this news, Leibniz finally dropped everything and hastened back from Vienna, intending to follow the Hanoverian court to London, where he hoped to reap dividends from the succession he had helped to arrange. Upon his arrival in Hanover on 14 September, however, he discovered to his dismay that the court had left without him three days earlier. To make matters worse, the new King George I explicitly prohibited him from crossing the Channel to join the court in London. First he had to finish the long-awaited Guelf history.

In the end, Leibniz died before he could fulfil this condition: he passed away in Hanover on the evening of 14 November 1716, with only his secretary and his coachman at his bedside, innumerable unfinished projects on his hands, and a mountain of unpublished papers on his desk. Yet, his happiness depended more on a vision for a better future than on taking stock of past success. 'Tranquillity is a step on the path toward stupidity,' he opined to Luise von Hohenzollern in 1705. 'One should always find something to do, to think, to plan, concerning ourselves for the community and for the individual, yet in such a way that we can rejoice if our wishes are fulfilled and not be saddened if they are not.' As he wrote in concluding his *Principles of Nature and of Grace* of 1714, 'our happiness will never consist, and should not consist, in complete satisfaction, where there is nothing left to desire, which would render our mind stupid, but in a perpetual progress to new pleasures and new perfections' (GP VI, 606).

Chapter 2
Characteristica universalis, logical calculus, and mathematics

How then did Leibniz propose to pursue his all-embracing programme of scientific advancement? What, in more detail, were the core projects which held his wide-ranging intellectual life together? From very early on, Leibniz nurtured the dream of developing an alphabet of human thoughts leading to the creation of a *characteristica universalis*, that is, a universal system of signs designed to eliminate the ambiguity of natural language. As well as benefiting theoretical sciences, this agreed system of signs would have constituted a formal language allowing the peaceful resolution of all manner of controversies: hence its immediate practical utility. This project, which from 1679 progressed into the development of a logical calculus, was meant to play a pivotal role in Leibniz's efforts toward reconciliation at a time of enormous religious, political, and intellectual upheaval. Over and above the provision of a means of universal and unambiguous communication, however, the *characteristica universalis* was conceived by Leibniz as a powerful tool of scientific discovery and judgement on the model of algebra. In short, one of Leibniz's most fundamental and long-standing projects was nothing less than a comprehensive programme of symbolic expression of relations amongst our thoughts, aimed at extending our reasoning power while minimizing error.

Signs and symbolic thought

This forward-looking programme of formalization of thought processes was grounded on the fundamental role Leibniz saw for signs and symbolic thought in human cognition. Following the teaching of the English philosopher Thomas Hobbes (1588–1679), Leibniz maintained that human thought always needed to be supported by sensible 'signs' such as written or spoken words, figures drawn on paper, and mental images. The problem was, Leibniz noted, that in many instances we assume that genuine concepts and ideas correspond to the signs employed in natural languages, when often this is not the case. Indeed, a great deal of our reasoning happens through 'blind thought' or 'symbolic thinking' in which we operate with signs without checking (or being able to check) whether these signs and symbols correspond to genuine concepts signifying possible beings. 'Blind' or 'symbolic thinking' is therefore both a strength and a weakness of human cognition: a strength because it immensely extends the putative grip of thought on reality and thereby its ability to operate; but a weakness because this grip is slippery. The signs and symbols that we employ in 'blind thought' instead of 'things' can easily hide a contradiction, when we overlook the fact that they do not ultimately 'latch on', as it were, to any possible being (A VI, 4, 587–8). In Leibniz's view, the creation of an artificial formal language (broadly conceived) provided the best solution to this problem via the establishment of a clear and unambiguous correspondence between signs and concepts. Following the usage of other authors, Leibniz used the term 'characters' (*characteres*) for the written signs which support thought and the term 'characteristic' for the discipline devoted to developing the system of signs which were meant to constitute the universal language.

The ideal of constructing a 'mental language' which could repair the Babel of natural languages was not new. It had been championed in the Middle Ages and rejuvenated in the early modern period in various attempts to create an artificial, universal language. Hobbes himself, in his *Computatio sive logica* of 1655, had promoted the idea that to think is to calculate. As already outlined in Leibniz's youthful *Dissertation on Combinatorial Art* (1666), this project required three main phases: (1) the systematic identification of simple concepts; (2) the choice of signs or characters to designate these concepts and to constitute a sort of universal alphabet of the artificial language; and (3) the development of a combinatorial method governing the combination of these concepts. Basically, phases one and three reworked the two procedures of analysis and synthesis fundamental both to traditional Aristotelian logic and to the demonstrative method established by the Ancient Greek mathematician Euclid (notwithstanding the important differences between analysis and synthesis in the logical as opposed to the mathematical tradition).

While situating his efforts within this established tradition, Leibniz was also aware that his conception of the *characteristica universalis* represented a crucial advance on previous attempts. In December 1678, he wrote to the mathematician Jean Gallois:

> I intend to make use of the Characteristic, of which I have spoken with you on occasion, and of which Algebra and Mathematics are merely samples. This Characteristic consists of a certain writing or language (since one who has one can have the other) which perfectly corresponds to the relations of our thoughts. This character will be completely different from those which have previously been projected because the most important thing has been overlooked: namely that the characters of this writing must assist discovery and judgement as in algebra and arithmetic. (A II, 1^2, 669)

The most revealing aspect of this letter to Gallois was the indication that Leibniz's *characteristica universalis* was pursuing a mathematization of logic as a tool for sound judgement and scientific discovery. Algebra and mathematics represented both the model to be followed and mere components of a broader project under which they were ultimately to be subsumed.

Logical calculus

In fact, one of the most important features of Leibniz's work towards the *characteristica* was the development of a logical calculus modelled on algebra. The groundwork for this logical calculus was laid in April 1679 in a group of essays in which Leibniz employed letters and numbers to indicate concepts and formulate propositions. In particular, he proposed to designate primitive concepts through prime numbers, to express complex concepts composed of primitive concepts as the product of prime numbers, and to translate into algebraic equations the four logical forms of categorical propositions represented by the 'square of opposition' (1. universal affirmative: every S is P; 2. universal negative: no S is P; 3. particular affirmative: some S is P; 4. particular negative: some S is not P; the logical relations amongst these four forms can be conveniently represented through a square diagram as shown in Figure 3).

For instance, translated into an algebraic equation, a universal affirmative proposition will be as follows:

$$\text{every S is P} \ldots\ldots S = y P$$

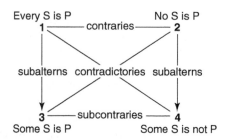

3. **Square of opposition.**

According to Leibniz, the universal affirmative proposition (e.g. 'every man is animal') should be read as affirming that every S (e.g. 'every man') is some species y of P (e.g. 'every man' is a rational species of 'animal', where 'rational' is expressed through the letter y). According to Leibniz's numerical interpretation, the universal affirmative proposition 'every man is animal', translated algebraically as $S = yP$, can also be formulated in numerical terms as '$6 = 2 \times 3$', where the complex concept 'man' is the product ('6') of simple concepts ('animal', 'rational') expressed by prime numbers ('2', '3'). By means of this new calculus, its proud inventor declared, all the modes and figures of the syllogism could be derived simply by applying the rules which govern numbers.

In logical texts written between 1678 and 1684, Leibniz outlined a *calculus universalis* focusing on the form of the argumentation and employing letters to designate terms (for instance: a = 'animal', b = 'rational', etc.). In 1686, in one of his most important logical writings (the *Generales inquisitiones de analysi notionum et veritatum*), Leibniz defined a notion of 'coincidence' which applied both to terms and to propositions: two terms (or two propositions) coincide (*coincidunt*) if they can be substituted for one another *salva veritate* ('preserving truth'), that is to say, if their substitution for one another does not alter the truth-value. He also distinguished in the logical calculus two fundamental relations: the relation of 'being included in' (expressed through the copula 'is' (*est*)) and the relation of 'coincidence' (expressed through the sign '='). Furthermore, he proposed a means of reducing categorical propositions to terms (e.g. the reduction of 'A is B' to 'the being B of A' or 'the B-ness of A'), and hypothetical propositions to categorical propositions. As he wrote in 1688, 'just as I reduce categorical propositions to simple terms affected by "*est*", I also reduce hypothetical propositions to categorical ones...For instance, I reduce this hypothetical proposition "If Peter is wise, Peter is just" to this categorical proposition: "to be wise Peter is to be just Peter". This way the same rules are valid for hypothetical propositions as are valid for categorical propositions' (A, VI, 992). A group of essays of 1686–7 contained further important developments, notably the introduction of a logical operation, labelled by Leibniz 'real addition' and symbolized by $\oplus$, in which concepts rather than numbers were added or 'composed'.

Remarkably, Leibniz's logical essays contained all the ingredients for developing what nowadays is called a 'Boolean algebra'—the symbolic system of mathematical logic which takes its name from the English mathematician George Boole (1815–64), and which provides the basis for the design of digital computer circuits. Leibniz's idea of expressing universal affirmative propositions as $S = yP$ will later be found also in Boole and other algebraists of logic of the end of the 19th century.

Dealing with disagreement and uncertainty

The role of the *characteristica* as a tool for the peaceful resolution of controversies through their formalization was in full view in a text of 1688. In it, the ideal of a reduction of reasoning to calculation was announced loud and clear:

> But to return to the expression of thoughts through characters, this is my opinion: it will hardly be possible *to end controversies* and impose silence on the *sects*, unless we reduce complex arguments to simple *calculations*, [and] terms of vague and uncertain significance to determinate *characters*....Once this has been done, whenever controversies arise, there will be no more need of a disputation between two philosophers than between two accountants. It will in fact suffice to take pen in hand, to sit at the abacus, and—having summoned, if one wishes, a friend—to say to one another: 'let us calculate' [*calculemus*]. (A VI, 4, 912–13)

The lovely image of two philosophers sitting down at the abacus with their friends to settle their differences by recasting their arguments in an unambiguous, formal language may seem far-fetched even for the average philosophical workshop, let alone the average clash between 'sects' of all sorts. It is therefore tempting to conclude that Leibniz was over-optimistic regarding human beings' willingness to follow reason. But *'calculemus'* was not the motto of an unworldly mathematician hopelessly overestimating the power of human reason. On the contrary, Leibniz was well aware of the limits of our intellect and often stressed that in areas crucial to human life such as medicine, jurisprudence, and religion, exact data were not (and often could not be) available. Nevertheless, he insisted, reason has a key role to play even in these areas. His *characteristica* was meant precisely to assist our judgement in uncertain situations thanks to its 'supreme exaltation, and *extremely efficient use of human reason by means of symbols and signs'* (A VI, 4, 913).

Rather than hoping for an improbable mathematical resolution of all uncertainties, Leibniz was envisaging a method for arriving at reasonable decisions in the many fields in which strict demonstrations were and would remain impossible. As a doctor of Law and a practising jurist, he was more accustomed than most to the idea that formalized and clearly regulated processes should and could be used for a peaceful and just resolution of controversies. His *characteristica universalis* was intended to help people to deal rationally with disagreement, to think through a problem, to weigh the reasons on both sides of an issue. When forced to operate (as it is so often the case) with conjectures rather than full data, Leibniz aimed to determine 'not only what is more plausible [*verisimilius*] but also what is *more secure* [*tutius*]'. For this purpose, he thought, we urgently needed a new part of logic: *'a part of logic*, so far virtually untouched, devoted to the estimation of degrees of probability; a steelyard of proofs, presumptions, conjectures, and clues' (A VI, 4, 914).

Leibniz was in fact acutely aware that much more needed to be done to create a logic of probability able to weigh reasons rather than counting them. Later in life, he even expressed the wish to give absolute priority in his own work to the development of this 'balance of reason'. As he wrote in 1697:

> It is often justly said that reasons should not be counted but weighed; but no one has yet given us a balance able to weigh the strength of reasons. It is one of the greatest deficiencies of our logic, from which we suffer even in the most serious and important matters of life, which regard justice, the tranquillity and wellbeing of the state, the health of human beings, and religion itself.... If God gives me more life and health, I will make this my principal concern. (A I, 13, 555)

Life was too short for Leibniz to develop a fully-fledged *characteristica*; and often runs too quickly for us to sit calmly 'at the abacus, and...say to one another: "let us calculate"'. Yet here again, the charge of otherworldliness is unjustified. Leibniz's *characteristica universalis* was not meant to supersede natural languages, which had, he recognized, vital roles in society at large and in our ability to communicate our views to one another. Leibniz treasured the concrete, vivid, emotive, and figurative language of poetry, drama, and narrative as more powerful in moving and motivating human beings than the non-emotive, rational language of philosophy or the abstract formalism of mathematics. He also acknowledged the classical insight that most people are more readily persuaded by rhetorical than by strictly logical means, and that in controversial issues one often needs to 'turn' the adversary, to make him see things from another perspective, rather than making him follow a chain of deductions.

Mathematics as a science of signs

The effectiveness of natural language notwithstanding, as a tool for the advancement of science, nothing could be more powerful,

Leibniz thought, than a science of signs broadly conceived. Algebra, arithmetic, and their applications to the physical world were prime examples of systems of symbols which, through rigorous rules for their manipulation, empowered discovery. Leibniz conceived his mathematical studies and his invention of the infinitesimal calculus (that is, an algorithm or sequence of rules for handling infinitesimal magnitudes) as part of this broader project.

Thus, while for Newton the calculus was basically a brilliant way to solve certain difficult mathematical problems, Leibniz regarded it as a sample of what he was hoping to achieve in logic for thought in general. Indeed, in a letter of 28 December 1675 to the secretary of the Royal Society, Henry Oldenburg, penned shortly after his discovery of the infinitesimal calculus, Leibniz linked his new 'algebra' to the *characteristica universalis*:

> This algebra (which we rightly praise so much) is only part of that general theory. Yet it ensures that we cannot err even if we wish to...To be sure, I recognise that whatever algebra supplies of this sort is the fruit of a superior science which I am accustomed to call either Combinatory or Characteristic, a science very different from what might at once spring to one's mind on hearing these words. (A III, 1, 331)

Leibniz's discovery of the calculus matured during his period in Paris (1672–6). In the 17th century, the mathematical study of motion and, in particular, the issue of how to measure curves, had become central to the revolutionary programme of providing mathematical accounts of natural phenomena ushered in by the new quantitative physics championed by Galileo Galilei (1564–1642). The calculation of the area of the surface intercepted by a plane curve (a problem known as the 'quadrature' of a curve), and the determination of a tangent to a given plane curve, leading to the calculation of the velocity of a body at a given instant, took centre stage in mathematical studies aimed at providing a general method of calculation applicable to all kinds of known curves and variable quantities.

One of the milestones on the road to the discovery of the calculus was Leibniz's generalization of the method of calculation introduced by Blaise Pascal (1623–62) in his work on the quadrature of the circle. As Leibniz triumphantly noted in a text of 1673, 'the whole thing depends on a right-angled triangle with infinitely small sides, which I am accustomed to call "characteristic", in similitude to which other triangles are constructed with assignable sides according to the properties of the figure. Then these similar triangles, when compared with the characteristic triangle, furnish many propositions for the study of the figure, through which curves of different kinds can be compared with one another' (A VII, 4, 597). (On the characteristic triangle, see Figure 4.)

Building on these results, a few months later Leibniz formulated a general 'transmutation theorem'. Further, Leibniz realized that tangents and quadratures were related to one another, showing that, in a sense, the problem of tangents is the inverse of that of quadratures. The arithmetical quadrature of the circle (achieved by Leibniz between the end of 1673 and the beginning of 1674) followed immediately from this theorem.

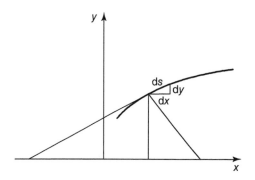

4. The characteristic triangle (ds, dy, dx).

In October 1675, Leibniz took the final steps leading to the invention of the differential and integral calculus, that is, respectively, the algorithm whose main application is the calculation of the slope of a tangent to a curve, and the algorithm whose main application is the calculation of the area under a curve. Writing some twenty years later to the Savilian Professor of Geometry at Oxford, the eminent mathematician John Wallis (1616–1703), he explained: 'consideration of the differences and of the sums in numerical series sparked my first flash of illumination when I realized that the differences corresponded to the tangents and the sums to the quadratures' (GM IV, 25). In fact, the recognition of the inverse relationship between the determination of tangents and of quadratures constituted one of his crucial discoveries. The invention of the differential calculus, providing a general method for the determination of tangents to curves, followed. The key 'general principle ... for measuring curvilinear figures', Leibniz later explained, is that '*a curvilinear figure must be considered equipollent to a polygon of infinite sides.* From this it follows that everything which can be demonstrated of such a polygon can be demonstrated of the curve, whether the number of sides is not taken into account at all, or whether it is made truer by assuming a greater number of sides, so that the error will always be less than any given error' (GM V, 126).

Leibniz conceived the division of a given curve represented in a Cartesian coordinate system into an infinity of infinitesimal intervals with extremes $s_1, s_2, s_3 \ldots$ The x-axis and y-axis of the Cartesian system were thus also divided into infinitesimal intervals with extremes, respectively,

$$x_1, x_2, x_3 \ldots$$

and

$$y_1, y_2, y_3 \ldots$$

The difference between two successive values of each of the variables s, x, y, that is, their 'differential' (d), was defined, respectively, as

$$\mathrm{d}x = x_{n+1} - x_n$$
$$\mathrm{d}y = y_{n+1} - y_n$$
$$\mathrm{d}s = s_{n+1} - s_n$$

The characteristic triangle has sides $\mathrm{d}x$, $\mathrm{d}y$, $\mathrm{d}s$ (see Figure 5). At this point, the tangent to the curve could be determined by treating the curve as equivalent to a polygon of infinite sides: 'To find the *tangent* means to draw a straight line connecting two points of the curve which have an infinitely small distance between them, that is to lengthen [one] side of the polygon with infinite angles which for us is equivalent to the curve. This infinitely small distance, however, can always be expressed through a known differential, such as $\mathrm{d}y$, or through a relation to it, that is through a known tangent' (GM V, 223).

In turn, working at the general solution to quadrature problems provided by the integral calculus, Leibniz equated the area of the

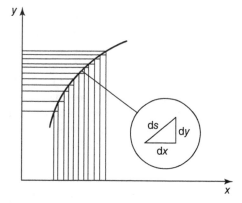

5. Differential representation of a curve.

surface subtended by a given curve to the sum of the areas $y dx$ of infinitely many strips (see Figure 5). In the margin of a manuscript of 21 November 1675, he introduced his distinctive symbol ∫—an elongated 's'—to indicate this sum (GB 161). Through his focus on the development of an efficient formal notation, Leibniz produced a very effective algorithm. This algorithm not only constituted a defining moment in the history of mathematics, with transformative consequences for modern physics. It also provided a striking example of Leibniz's general point regarding the power of discovery afforded by rigorous formal systems in which symbolic thinking immensely extends our grasp of nature and our ability to operate while minimizing error.

Leibniz waited nine years before publishing his great discovery of 1675. In 1684 the *Nova methodus pro maximis et minimis* appeared in one of the learned journals of the time, the *Acta eruditorum* (see Figure 6). Newton, however, waited even longer to disclose publicly his own 'method of fluxions', discovered roughly a decade prior to Leibniz for the solution of the same set of problems. Study of the notes in which Newton and Leibniz developed their methods has now provided conclusive evidence that their discoveries were independent of one another. This fact did not prevent the eruption of one of the most heated priority disputes in the entire history of science and mathematics. Besides the original issue, the controversy metastasized, extending to a national clash between the English and German scientific communities, and involving lesser personages who had their own agendas and conducted their own wars by proxy. One cannot help wishing that the idyllic picture of two philosophers at the abacus resolving their differences by calculation had applied at least to the two greatest early modern mathematicians. Unfortunately, however, their friends were not the sort to sit peacefully around Leibniz's abacus.

One of Newton's friends, John Keill (1671–1721), used the official periodical of the Royal Society, the *Philosophical Transactions* for

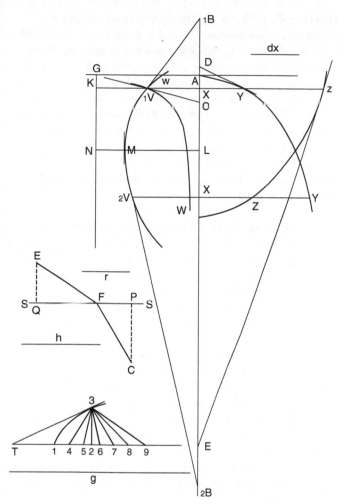

Characterística universalis, logical calculus, and mathematics

6. Diagrams placed opposite the first page of Leibniz's *Nova methodus pro maximis et minimis* in the *Acta eruditorum* of October 1684 to illustrate his new method of differentiation.

1708 (actually published in 1710), openly to accuse Leibniz of having plagiarized Newton's method of fluxions. Leibniz, himself a fellow of the Royal Society, complained to its secretary, Hans Sloane, demanding an apology. Sloane then turned for advice to the Society's President—an unobjectionable move, were it not for the fact that the President was none other than Newton himself. Instead of being asked to apologize, Keill was invited to present his account of the alleged plagiarism in a report which was read at the Royal Society meeting on 24 May 1711 and then copied to Leibniz. In December 1711 Leibniz sent a letter to Sloane in his defence. As he saw it, Newton had a right to claim independent discovery, but so did he.

To settle the issue, the Royal Society ordered a review of the available documents and a final report. The investigation resulted in a report, framing Leibniz as a plagiarist, which sailed through the Society's peer review process with exemplary celerity, sped on its way not least by the fact that it was written by Newton himself. Publication of the damning report and supportive documents were ordered, resulting in a volume that Newton also took upon himself to edit and annotate. The *Commercium epistolicum D. Johannis Collins et aliorum de analysi promota* appeared in London in January 1713 (bearing 1712 as publication year). Leibniz did not see it for himself until September 1714, when he returned from Vienna to Hanover; but reports of its contents had already reached him. Incensed, he wrote anonymously, in the summer of 1713, a short leaflet (christened by Newton *Charta volans* or 'flying paper') intended for broad circulation. In it, the anonymous author supported the suggestion that, if anything, Newton had plagiarized Leibniz's calculus, not the other way round.

In February 1716, Leibniz offered Newton an olive branch via his friend Caroline, now Princess of Wales: he would not mind acknowledging Newton's priority as long as the independence of his own discovery was recognized. But nothing came of this belated peace offering. On 10 December 1716, one of Newton's supporters visiting Hanover wrote to the English mathematician, 'Mr Leibniz is dead and the dispute is finished,' adding that the German savant had 'laboured all his life to invent machines which did not work'. This prognostication of the end of the dispute proved as inaccurate as this assessment of Leibniz's life's work was uncharitable. By 1716, the calculus controversy had taken on a life of its own, which long survived the death of one of its two main protagonists.

None of Leibniz's other momentous mathematical advances attracted similar attention during his lifetime, but neither did they vex him to the same degree. One of these was his invention of determinants and the discovery of their properties. Determinants are used to solve systems of linear equations through the elimination of the unknowns, for instance,

$$a_{10} + a_{11}x + a_{12}y = 0$$

and

$$a_{20} + a_{21}x + a_{22}y = 0$$

in which x and y are the unknowns. In a manuscript of 22 January 1684, Leibniz reached the general rule for the solution of systems of linear equations through determinants, now known as 'Cramer's rule' after the mathematician Gabriel Cramer (1704–52) who formulated the rule later on. It is worth noting that Leibniz's interest in determinants was driven, once again, by the broader project of the *characteristica universalis*, that is, by his comprehensive plan for creating efficient systems of signs or symbols, with rules for their combination.

Part of this broader project was also the *characteristica geometrica* or *analysis situs* (analysis of situation). The key idea behind this new type of geometry was to express situation, angles, and

7. One of Leibniz's manuscripts on his novel binary arithmetic or dyadic.

movements by symbols instead of figures, and then to operate with them in an efficient and clear way (A III, 2, 851–3). Leibniz's novel geometrical science was mentioned as early as 1678 in a letter to Gallois. By August 1679, a first essay entitled *Characteristica geometrica* was ready (GM V, 141–68). In September 1679, Leibniz sent a paper to Christiaan Huygens outlining his 'new characteristic' (A III, 2, N. 347). In a letter of October 1679, he described it enthusiastically to Huygens as 'an overture that, in its field, should take us as far as algebra in its respective field' (A III, 2, 875). Regrettably, Huygens dismissed Leibniz's attempts as 'wishful thinking', noting that the sample demonstrations employing symbols instead of geometrical figures went over well-known results (A III, 2, 888–9). The failure of Leibniz's erstwhile mathematical mentor to appreciate the ground-breaking potential of the *analysis situs* did not discourage his former pupil. It was not until well after Leibniz's death, however, that the significance of his discovery began to be appreciated as a transformative shift from a geometry of figures to a geometry of space.

A similar fate met another amazingly forward-looking pet project, namely, his development of a binary arithmetic or dyadic (Figure 7). Belittled at the time by the Parisian Académie Royale des Sciences as less fruitful than the well-oiled decimal system, Leibniz's dyadic had to wait until the 20th century to see its potential vindicated by the advent of the brave new world of computers. As for the innovative calculating machine at which Leibniz laboured for most of his life, its pioneering features were repeatedly played down as scarcely improving on more basic, tried-and-tested models.

Chapter 3
Encyclopaedia, *scientia generalis*, and the academies of sciences

The *characteristica universalis* was intended in turn to provide the key instrument of a *scientia generalis* (general science) leading to the development of a new kind of encyclopaedia of all the sciences. In the decades preceding Leibniz's birth, the encyclopaedic ideal of presenting all available knowledge in an easily mastered format had become the central aspiration of a group of scholars now known in the Leibniz literature as the 'Herborn Encyclopaedists'. Herborn was a small town in the tiny German county of Nassau-Dillenburg, one of an archipelago of Reformed statelets which, together with several hundred semi-autonomous political entities of various Christian denominations, formed the Holy Roman Empire. The need to provide quasi-university education within the limits of a small state had driven Nassau-Dillenburg to focus on pedagogically effective instruction methods shunned by larger schools and universities, and the capacity of the Herborn Academy to punch way above its weight was an internationally recognized testimony to the success of those methods. Building on the system of basic topics or commonplaces (*loci*) developed by the French humanist pedagogue Petrus Ramus (or Pierre de la Ramée, 1515–72) in his attempts to simplify Aristotelian logic, Herborn's teachers had reinserted Aristotelian content into a Ramist framework, while also developing a combinatorial logic inspired by the Catalan thinker Ramón Lull (1235–1315). These developments cumulated in the seven-volume *Encyclopaedia*, compiled by

Johann Heinrich Alsted (1588–1638) and published in Herborn in 1630: the greatest encyclopaedia of its era and the first of a long line of major works to bear that title. Scarcely less important for Leibniz was the work of Alsted's son-in-law, Johann Heinrich Bisterfeld (*c.*1605–1655), who further refined the epistemological and metaphysical conceptions at the heart of Alsted's magnum opus. Bisterfeld's metaphysics provided a profound source of inspiration for the young Leibniz and Alsted's work the major departure point of his own encyclopaedic project.

The 'demonstrative' encyclopaedia and the *scientia generalis*

Leibniz's initial plan was to undertake the collection and systematic reorganization of all available knowledge into an updated version of Alsted's universal encyclopaedia of all the arts and sciences. But even this immense task was merely a first step toward the development of an altogether new kind of work: an 'inventive' or 'demonstrative' encyclopaedia. It was not sufficient, in his view, to review, collect, and make available in a concise and convenient format the chief discoveries of each field of knowledge. The ongoing advancement of knowledge required the systematic identification of the principles and methods from which arts and sciences had developed in the past and could develop in the future. To this end, it was first necessary to review existing results and practices in order to arrive at a 'general science' (*scientia generalis*) of the principles, foundations, and methods of all the sciences. The 'inventive' or 'demonstrative' encyclopaedia could then have been expounded as a result of the systematic and cross-disciplinary application of relevant principles and methods. This grand vision of the advancement of all the sciences was grounded in a conviction of the unity and systematicity of knowledge, in which some fundamental truths were shared by all sciences, and the results in one field of knowledge could serve as points of departure for further discoveries in other fields. 'It does not make much difference how you divide the sciences',

Leibniz wrote, 'for they are a continuous body, like the ocean' (A VI, 4, 527).

His descriptions of the object and tasks of the *scientia generalis* varied somewhat from one period and text to another. Sometimes he assigned a narrower meaning to this project by depicting the *scientia generalis* as a science of method, that is, a science which 'teaches all other sciences how to discover and demonstrate from sufficient data' (A VI, 4, 370). In this narrower sense, the *scientia generalis* could be seen as corresponding to logic in its classical sense of the study and development of the tools of valid reasoning. In a broader sense, however, the *scientia generalis* would have embraced the principles and foundations of all the sciences, leading to the establishment of the elements of the whole encyclopaedia and to the investigation of the highest good.

The goal of the whole enterprise was therefore not theoretical but practical: namely, the achievement of 'discoveries for the benefit of public happiness' (A VI, 4, 525). The *scientia generalis*, with its direct remit of promoting the 'instauration and advancement of the sciences' (A VI, 4, 525), was in fact ultimately directed at the search for wisdom, defined as 'the science of happiness'. As Leibniz explained in 1678–9:

> In my view, wisdom is nothing other than the science of happiness, and true learning, considered as a preparation for wisdom, is the habit of a soul most plentifully supplied with the knowledge of how to live well and happily.... From this it follows that it is in the interest of the happiness of humankind that there be brought together a certain *encyclopaedia* or orderly collection of truths, sufficient (as far as possible) for the deduction of all useful things. And this will be like a public treasury to which could be added all remarkable [subsequent] discoveries and observations. But since [this encyclopaedia] will be of the most massive bulk, especially regarding matters of civil and natural history, in the meantime a certain *Scientia Generalis* is needed containing the first principles of reason and experience. (A VI, 4, 137–8)

Academies of sciences

Leibniz was fully aware that this all-embracing project ought to be a collaborative enterprise undertaken on a scale that only an enlightened ruler could support. His indefatigable attempts to found academies (or rather 'societies') of sciences were therefore directly connected to his encyclopaedic ambitions.

His efforts met with at least partial success in Berlin, where a Society of Sciences was established by Friedrich III of Brandenburg in 1700. Leibniz, as first president, chose 'Theoria cum Praxi' as the Society's motto, epitomizing his lifelong ideal of wedding theory to praxis. The division of the Society into four main departments—devoted to natural and medical sciences; to mathematical, astronomical, and mechanical sciences; to Germanic and historical studies; and to literary-historical and oriental studies—attempted to cover systematically the main fields in the arts and sciences. In 1710 Leibniz was able to see through the press the first volume of the official periodical publication of the Society, the *Miscellanea Berolinensia ad incrementum scientiarum*. The volume included sixty papers organized into three sections: the first, on literary matters; the second, on physical and medical research; and the third, on mathematical and mechanical enquiries. One-fifth of the papers were contributed by Leibniz himself, and covered a range of topics across all three sections: from his studies of fossils to the aurora borealis; from the infinitesimal calculus to the description and illustration of his calculating machine; from the history of the invention of phosphorus to an essay on the origin of peoples traced through the study of languages, and offering one of the most significant systematizations of his thought as a linguist and philosopher of language.

Plans for other scientific organizations punctuated his life until the very end—some equally ambitious in their aim of fostering all the arts and sciences; others on a smaller scale and with a more specific focus. They were often accompanied by detailed proposals on how to secure financial support, ranging from taxation on the consumption of tobacco to the farming of silkworms. This constant flow of memos and schemes was not matched by the sober reports on outcomes which even early modern funding bodies seem to have expected. Yet, the perseverance with which he continued to put forward creative proposals in the face of recurrent frustration testifies to the centrality of these collaborative scientific endeavours to his intellectual programme. A renewed attempt to secure patronage, addressed on 16 January 1712 to Peter the Great, beautifully summarized a lifelong aspiration centred on the advancement of science as the key instrument for the improvement of the human condition:

Although I have very frequently been employed in public affairs and also in the judiciary system and am consulted on such matters by great princes on an ongoing basis, I nevertheless regard the arts and sciences as a higher calling, since through them the glory of God and the best interests of the whole of the human race are continuously promoted. For in the sciences and the knowledge of nature and art, the wonders of God, his power, wisdom, and goodness are especially manifest; and the arts and sciences are also the true treasury of the human race, through which art masters nature and civilised peoples are distinguished from barbarian ones. For these reasons I have loved and pursued science since my youth. . . . The one thing I have been lacking is a leading prince who adequately embraced this cause. . . . I am not a man devoted solely to his native country or to one particular nation: on the contrary, I pursue the interests of the whole human race because I regard heaven as my fatherland and all well-meaning people as its fellow citizens. . . . To this aim, for a long time I have been conducting a voluminous correspondence in Europe and even as far as China; and for many years I have not only been a fellow of the French and English Royal Societies but also direct as president the Royal Prussian Society of Sciences. (Guerrier, 206–8)

Chapter 4
Possible worlds and fundamental principles

Leibniz's commitment to the *characteristica universalis* and the *scientia generalis* ultimately rested on his conviction that logic is a mirror of the structure of reality. In his view, the principles which govern thought were also the principles that govern reality. In turn, reality meant for Leibniz first and foremost God, the eternal and infinite Being encompassing all perfections. It is from him and his eternal thoughts that the story of the world in which we find ourselves begins. Logic therefore led via metaphysics to philosophical theology. Against the backdrop of Leibniz's overarching projects, we can now follow him on a very short journey through possible worlds, the thoughts of God, and the actual world springing from them, focusing on some of his trademark philosophical views.

Identity/non-contradiction, possible beings, and possible worlds

According to Leibniz, the first, fundamental principle which governs both the ideal order (the sphere of thought, studied by logic) and the real order (reality, studied by metaphysics) is the principle of non-contradiction: for any proposition 'p', 'p' and 'not-p' cannot both be true at the same time in the same respect. In other words, it cannot both be true at time t that I am such and such (whatever that is) and that I am *not* such and such. It cannot both be true that I am, right now, wearing my glasses on my nose and *not* wearing my glasses on my nose, even if I may not know which one of these alternatives is true (say, by mistake, I am wearing your glasses, which happen to be the same model, colour, and strength as mine). For Leibniz, this is equivalent to the principle of identity, that is, the principle that A is A—everything is what it is and not what it is not (these glasses are these glasses, namely, either they are my glasses or they are not my glasses; either their frame is made of metal or it is not made of metal; either they were bought in London at time *t* or they were not bought in London at time *t*, irrespective of whether we know which one of these alternatives is true). As Leibniz puts it:

> *Primary truths* are those which assert the same [thing] of itself, or deny the opposite of its opposite. As *A is A*, or *A is not not-A*. If it is true that *A is B*, it is false that *A is not B*, or that *A is not-B*. Also: *every thing is what it is. Each thing is like itself, or equal to itself. Nothing is greater or less than itself*—and others of this sort which, though they may have their own grades of priority, can all be included under the name of 'identities'. (A VI, 4, 1644)

Only beings which are possible, that is, beings which do not imply contradiction, can be thought or conceived. We can of course think the words 'square circle' but thinking the words does not mean conceiving a being which is both square and circle. We can only conceive a square *and* a circle, not a square circle. And this is not just a matter of having, as human beings, a limited intellect. According to Leibniz, not even God could have the idea (or contemplate the essence) of a square circle because there is not, and there cannot be, such an idea. There cannot be a geometrical figure which is both square and round because 'A' (a circle) cannot be 'not-A' (not a circle). In brief, a square circle is not a possible being.

Moreover, to be a possible being, that is, to be a being which can be thought or conceived, is not yet to be a *really existing* being, that is, a being having some extra-mental existence. For Leibniz, possible but non-existing individuals are just thoughts or 'ideas'. They have no other kind of existence than mental existence. As such they obviously need a mind thinking them or they would not have any existence at all. Hence, a mind is needed to ground whatever lower degree of reality there is in beings which are possible but do not really exist. For instance, I could surely think of my imaginary twin sister being in Paris right now while I am stuck in London. Even if the imaginary twin sister I never had is nowhere to be found in Paris, still she has some kind of reality or existence as a thought in my mind—a lower degree of reality which would not be there at all if absolutely no one thought of her.

It must be stressed, however, that for Leibniz the mental reality of non-existing possible beings does not ultimately depend on their being thought by individual human minds. One could take the view (as Leibniz did) that there are some truths (such as mathematical truths) which exist in some sense, or have some 'reality' of their own, even if no human being ever thinks of them. Even if no human being ever thought of it, 'there would still exist the impossibility of a square larger than an isoperimetric circle [that is, a circle of equal perimeter]' (A VI, 4, 17). Or, as Leibniz wrote in another note of 1677:

> It is true, and even necessary, that the circle is the largest of isoperimetric figures. Even if no circle really existed. Likewise even if neither I nor you nor anyone else ever existed.... Since, therefore, this truth does not depend on our thought, there must be something real in it. (A VI, 4, 18)

The question is in what sense these truths exist, and which kind of reality they have. Truths, for Leibniz, are not the sort of beings which can 'subsist' on their own without some entities which make them true (e.g. squares and circles, existing at least as objects of logic), and without a recognition of some fact about these entities and their relations. According to him, 'all true predication has some foundation in the nature of things' (DM 8; A VI, 4, 1540), that is, whatever is true is true in virtue of some entity which makes it true. Moreover, truths expressing some relations about entities 'result' by the thinking together of these entities (say, some mind thinking together a square and a circle).

If one grants that there would be eternal, necessary truths about mathematical objects even if no human mind ever thought of them, or even if there were no really existing beings instantiating some geometrical figure, one has to conclude by this reasoning that there must be some other, superior mind which grounds their reality by eternally thinking these objects and their relations. Following a long tradition inspired by Augustine's Christianized version of Platonism, Leibniz concluded that this mind is the mind of God. God eternally thinks all possible beings, and this is what ultimately grounds the reality of those possible beings which exist only as thoughts or ideas. As Leibniz explained in the *Monadology*:

> It is also true that in God is the source not only of existences but also of essences, insofar as they are real, or of what there is of real in possibility. This is because the Understanding of God is the realm of eternal truths, or of the ideas on which they depend, and because without him there would be nothing real in the possibilities—not only nothing existent, but also nothing possible. For if there is some reality in essences or possibilities, or indeed in eternal truths, this reality must be founded on something existent and actual; and consequently on the existence of the Necessary Being in whom essence includes existence, or in whom it is sufficient to be possible in order to be actual. (GP VI, 614)

God's infinite intellect thus embraces the ideas of all possible beings, that is, all beings which can be thought because they do not imply a contradiction. In God's mind, possible beings are in turn organized in worlds in which they are possible *together*. It is logically possible (that is, it does not in itself imply a contradiction) that the bullets fired at President Kennedy at 12:30 p.m. on 22 November 1963 in Dealey Plaza, Dallas, Texas, might have missed their target. But it is *not* logically possible that what actually happens in one given world (e.g. that President Kennedy *was* struck by those bullets at that time and place) also does *not* happen in this same world (i.e. that President Kennedy was *not* struck by those same bullets at that same time and place). These two events, although in themselves both logically possible, are not *com-possible* in any given world. They therefore belong to two different possible worlds.

The principle of sufficient reason, truths of reason, and truths of fact

In each of these possible worlds, 'nothing is without a reason' (A VI, 1, 494; A VI, 2, 483): everything has 'a sufficient reason, why it should be thus and not otherwise, even though most often these reasons cannot be known to us' (*Monadology*, § 32; GP VI, 612). This is Leibniz's principle of sufficient reason—the second, fundamental principle governing both logic and metaphysics, the ideal order and the real order.

To endorse the principle of sufficient reason is to endorse the thoroughgoing intelligibility—that is, rationality—of reality. For Leibniz, 'no fact could be found to be true or existing, and no proposition to be true, unless there is a sufficient reason' (*Monadology*, § 32; GP VI, 612). In other words, there are no brute, or unexplainable, facts; there is no ultimate, ground-floor irrationality from which things spring randomly. No matter how unreasonable, inexplicable, or chancy things may appear to us, there is an explanation of why they are the way they are. All facts are the result of an infinite chain of reasons which connects everything with everything else, even if, due to this very infinity, it is not possible for us to follow or discover these reasons.

In turn, the principle of identity/non-contradiction and the principle of sufficient reason ground, respectively, two kinds of truths: truths of reason and truths of fact. Truths of reason 'are necessary and their opposite is impossible' (GP VI, 612). That is, a truth of reason is a truth the opposite of which implies contradiction. For instance, truths of reason or necessary truths include mathematical propositions such as 'the sum of the internal angles of a Euclidean triangle is 180 degrees'. In the context of Euclidean geometry, to say that the sum of the internal angles of a triangle can be more than 180 degrees is to say that a 'triangle' can be 'not-a-triangle', or that 'A' is 'not-A'. This is impossible, that is, it implies contradiction.

On the other hand, truths of fact are not necessary: their opposite is false but not contradictory. If, as a matter of fact, I am wearing a coat, it is false that I am not wearing a coat, even though it would not have been contradictory (that is, it would not have been logically impossible) for me not to wear a coat right now. Thus, 'truths of fact are contingent' (GP VI, 612), in the sense that they are true by virtue of the way things in fact are and not by logical necessity. The way things in fact are, however, is a result of the chain of reasons which explains why they are thus and not otherwise. For example, I am wearing a coat because it is cold in my room; it is cold in my room because there is no central heating and it is winter; there is no central heating because the boiler is old and has broken down, and it is winter because of the inclination of the axis of the earth in relation to the sun at this time of the year in the northern hemisphere, etc., etc.—the process of giving sufficient reasons of the truth of fact that I am wearing a coat rapidly escalates to cosmic dimensions.

Universal harmony

If one stretches these reasons to infinity in every direction, Leibniz's claim that everything is linked to everything else begins to appear less outlandish. If one holds, on the basis of the principle of sufficient reason, that everything unfolds according to a rational order in which things are the way they are because of an infinitely complex chain of reasons connecting the past, present, and future of everything, to the past, present, and future of everything else, then it makes sense also to claim (as Leibniz did) that each change in any individual is reflected in the whole universe. In his phrase, there is a 'universal harmony' connecting everything to everything. The thesis of the ancient Greek physician Hippocrates (*c.*460–*c.*375 BCE) that in the human body everything 'breathes together' (*sympnoia panta*) can be extended to the whole world.

Harmony is defined by Leibniz as 'similitude in variety', 'diversity compensated by identity', or 'unity in multiplicity': 'Since [harmony and discord] consist *in the proportion of identity to diversity*, harmony is, in fact, unity in multiplicity; it is greatest when a plurality of elements... are resolved into the greatest concordance' (A VI, 3, 116, 122). A possible world is a 'unity in multiplicity'. It is a combination of the possible beings which are possible together (or com-possible), that is, the combination of the 'diversity' and 'variety' of things which can 'harmonize' or be 'unified' in the same world.

The thesis that everything is connected to everything else also means, for Leibniz, that each individual 'mirrors', 'reflects', or better, 'expresses' the entire universe from its point of view (however imperfect and confused this 'expressing' or 'mirroring' might be):

> *Every individual substance involves in its perfect notion the entire universe*, and everything existing in it, past, present and future.... Indeed, *all created individual substances are diverse expressions of the same universe*, and of the same universal cause, namely God; but they vary in the perfection of this expression, like the different representations or projective drawings of the same town from different points of view. (A VI, 4, 1646)

Extrinsic denominations, identity of indiscernibles, and relations

Moreover, according to Leibniz, '*there are no purely extrinsic denominations*, which have absolutely no foundation in the very thing denominated' (A VI, 4, 1645). By 'denomination' is meant a property or attribute of a subject which can be described in terms of its being 'denominated' by this property. Traditionally, there are two types of denomination: intrinsic denomination and extrinsic denomination. Intrinsic denominations are properties which are intrinsic to a being taken in itself. For example: 'the wall is made of bricks' is an intrinsic denomination since it describes a property of the wall taken in itself. Had the wall been made of stones, it would have been different. On the contrary, 'the wall is seen' is an extrinsic denomination. 'Being seen' does not seem to be making any difference to the wall. It is a denomination or a description of the wall which does not capture one of the wall's own 'internal' or intrinsic properties, but something 'external' or extrinsic to it. An extrinsic denomination is thus a kind of relational property. It requires the simultaneous consideration of two (or more) subjects with their properties and results from this comparison. For example, it requires a wall and a person seeing it, and their simultaneous consideration.

In Leibniz's time, the standard Scholastic view was that extrinsic denominations are founded in intrinsic properties of the subjects. Although 'being seen' is an extrinsic denomination of the wall, the wall must have some intrinsic properties (e.g. it must be made of a non-transparent material) in order to be seen. It was also normally held that a change in the intrinsic properties of only one of the subjects involved in a relation (in this case, the wall and the person seeing it) would have as a result a change of relation without necessarily implying also a change in the properties of the other subject. Take, for instance, two white men—Socrates and Plato. They are similar insofar as they are both white. Now (to use Leibniz's own example) cover Socrates in black ink. Their relation of similarity changes to a relation of dissimilarity, but only the intrinsic properties of one of the subjects seem to have changed. Nothing seems to have changed in Plato, who is still white, although the relation between Socrates and Plato has changed.

Leibniz departs from this view in maintaining that a change in relation between two subjects is necessarily accompanied by a change in the intrinsic properties of not only these two related subjects, but of *all* the individuals in the world. This is ultimately a consequence of the principle of identity/non-contradiction. If a property of a thing, no matter how seemingly 'external' to that thing, were in fact totally extraneous or 'extrinsic' to the thing to which this property is truly attributed (or of which this property is truly 'predicated') that thing would be at the same time itself and not-itself (namely, it would be also something totally extraneous to what that thing is). But this would be against the principle of identity/non-contradiction according to which 'A' is not 'not-A'. Hence, every property which is truly predicated of a thing must have some foundation in the intrinsic properties of that thing.

From this it follows that there are no purely extrinsic denominations, that is, properties which are truly attributed to a thing but which make no difference at all to *what that thing is*. Leibniz's claim that there is a universal harmony connecting everything to everything else should be interpreted as a claim that each of these connections is grounded in the intrinsic properties of each individual belonging to a given possible world. Thus, to use another Leibnizian example, the Emperor of China as known by me is *intrinsically* different from the Emperor of China as not known by me.

If this is the case, another radical claim by Leibniz begins to fit into the puzzle as a key element of an internally coherent net of radical views. This is Leibniz's principle of identity of indiscernibles. This principle, according to Leibniz, follows straightforwardly from the principle of sufficient reason and is also entailed by Leibniz's doctrine that no denominations are purely extrinsic:

> *there cannot be in nature two individual things which differ in number alone.* For it certainly must be possible to give a reason why they are diverse, which must be sought from some difference in them.... two perfectly similar eggs, or two perfectly similar leaves or blades of grass in a garden, will never be found. (A VI, 4, 1645)

According to Leibniz, the common illusion that there may be two perfectly similar individuals which differ in number alone (say, two perfectly similar grains of salt in the whole wide world) arises from the incompleteness of our abstract notions, and by our taking what our senses or our intellect manage to discriminate as an accurate account of how things are. That is, when we think of a grain of salt we 'abstract' or consider only a finite number of features of concrete grains of salt, from which we form the notion of 'grain of salt'. This is, however, an incomplete notion which does not capture and cannot capture the infinite complexity of a concrete, really existing individual, connected as it is (as all other concrete individuals) to everything else, and given that this makes an intrinsic difference to what this individual is. Such an infinite complexity is not something of which our intellect—let alone our senses—can give us an adequate account. Numerically distinct but *intrinsically* perfectly similar individuals are not possible within the same possible world. Nor can there be two indiscernible individuals each belonging to different possible worlds, since each individual mirrors the entire universe of which it is part.

The main pieces needed to appreciate the overall shape of Leibniz's theory of relations are now in place. According to Leibniz, relations are mental entities with foundations in the properties of things. More precisely, relations are not 'things' but 'truths' about things which 'result' when two or more subjects (and their properties) are thought together. In the *New Essays*, commenting on John Locke's way of distinguishing the main kinds of 'objects of our thought', Leibniz declares:

This division of the objects of our thought into substances, modes, and relations is rather agreeable to me. I believe that qualities are nothing else than modifications of substances and the understanding adds relations to them. (A VI, 6, 145)

Thus, according to this taxonomy, there are individual beings (substances), their modifications (that is, their qualities or properties), and relations which the mind 'adds' when thinking together these individual beings with their qualities. Any 'reality' relations have depends entirely on the reality of the individual beings in which relations are founded, and on the mind which thinks these individual beings together. Without a mind grasping these individuals together, there would still be the *foundations* of relations (that is, the intrinsic properties of the individual beings on which relations are grounded), but not relations as such. For there to be the relation 'Francesca is taller than Sophia', there must be Francesca and Sophia, having, respectively, the intrinsic property of being, say, 159 cm tall and 140 cm tall, and a mind considering these two individuals together and seeing the truth 'Francesca is taller than Sophia' resulting from this co-cogitation (to use Leibniz's newly forged notion of *concogitabilitas*).

Leibniz also claims, however, that relations 'have a reality beyond our intelligence'. The reality of relations is ultimately grounded not in individual human minds thinking them, but in the Divine mind eternally thinking the ideas or essences of all possible individuals, organized into the possible worlds in which these possible individuals are com-possible. In contemplating all possible individuals, God also thinks all truths which can be predicated of these individuals and all relations which result, as second-level truths, from these individuals (with their properties) thought together. In brief, God is the 'root' of every reality, including whatever degree of reality there is in truths, relations, and possibilities (A VI, 4, 1618).

Chapter 5
Complete-concept theory, theory of truth, and theory of knowledge

Leibniz maintains that the principle of identity/non-contradiction (A is A; A is not not-A; each thing is what it is and not what it is not) implies that every true predication—that is, every property which is attributed with truth to a certain thing—must have some foundation in the nature of that thing (DM 8). Otherwise A would not be A. If everything is connected to everything and all denominations are intrinsic (that is, have some foundation in the nature of a thing), what does this mean for an individual in a possible world?

Individual substances, complete concepts, and truth

In the *Discourse on Metaphysics* (1686), Leibniz defines an 'individual substance' as the single subject to which several predicates are attributed while 'this subject is not attributed to any other' (DM 8; A VI, 4, 1540). For instance, Alexander the Great is the subject of a number of predicates (e.g. 'being a king') but is not predicated of any other subject. This was a traditional, broadly Aristotelian view of what counts as a 'substance', namely, what counts as the sort of individual being that is metaphysically fundamental or primary. However, Leibniz immediately adds, 'this is not enough' since it is necessary 'to consider what it is to be truly attributed to a certain subject':

> Now, it is firmly established that all true predication has some basis in the nature of things, and when a proposition is not identical, that is, when the predicate is not contained expressly in the subject, it must be contained in it virtually. This is what philosophers call *in-esse*. Thus the subject term must always include the predicate term, in such a way that a man who understood the notion of the subject perfectly would also judge that the predicate belongs to it. (DM 8; A VI, 4, 1540)

In other words, all true propositions can be reduced, in principle, to identities. To say that Sophia Loren is Italian is to say that 'being Italian' is part of what it is to be Sophia Loren, it is part of her identity, in brief, it is part of articulating that A (Sophia Loren) is A (Sophia Loren). For anything I say about Sophia Loren to be true, that something must be included in the concept or the notion of Sophia Loren; it must be a way in which I progressively articulate (however imperfectly) what are the properties, attributes, or qualities which, taken together, make up what it is to be Sophia Loren. Thus the proposition 'Sophia Loren is German' is false because 'to be German' is not part of who Sophia Loren is and therefore is not included in the concept or notion of Sophia Loren.

Suppose that we now discover that Sophia Loren is Sofia Scicolone. If all that is true of Sophia Loren is true of Sofia Scicolone, then they are identical, that is, they are, really, the same individual substance which is identified by all the past, present, and future attributes or properties which are true of that individual substance. If instead we were to discover that, say, Sophia Loren was born in Naples but Sofia Scicolone was born in Rome, we would conclude that they are not, after all, the same woman. In brief, whatever name one attaches to an individual substance A, A is A. The identity of this A (who this A is) is determined by *all* the properties past, present, and future which are true of A: in this case, the properties 'being born Sofia Scicolone' and 'taking on Sophia Loren as a stage name' would be included in the concept of A.

If I possessed a really comprehensive knowledge of A, I would know all the properties which belong to A, that is, I would be able to judge whether or not a certain predicate is included in the notion of a certain subject. Moreover, I would also know what will happen to A in the future, that is, I would know a priori, or independently of experience, whether a certain property will belong to A, because that property would also be part of what A (a certain individual substance) is.

Leibniz expresses all this through his complete-concept theory, which maintains that there is a 'complete' concept corresponding to each individual substance, a concept which includes all the predicates (past, present, and future) which can be predicated with truth of that individual substance. Given Leibniz's principle of sufficient reason, his doctrine of universal harmony, and his denial of purely extrinsic denominations, it is easy to see that these predicates, in turn, will ultimately mirror all that happens in the universe since each property or attribute will be the result of an infinite chain of reasons which explain why something is thus and not otherwise, connecting everything with everything else:

> That being so, we can say that it is the nature of an individual substance, or of a complete being, to have a notion so complete that it is sufficient to contain, and make deducible from it, all the predicates of the subject to which this notion is attributed.... God, seeing the individual notion or *haecceitas* of Alexander, sees in it at the same time the foundation and the reason of all the predicates which can truly be stated of him, as for example, that he will vanquish Darius and Porus, to the point of knowing *a priori* (and not by experience) whether he died a natural death or died by poison, something we can know only through history. Therefore, when one considers properly the connection of things, one can say that there are in the soul of Alexander, from all time, vestiges of all that has happened to him,

and marks of everything that will happen to him, and even traces of
everything that happens in the universe—even though it is proper
only to God to recognize them all. (DM 8; A VI, 4, 1540–1)

The other face of the coin of Leibniz's complete-concept theory is his
conception of truth as the inclusion of the predicate in the subject:
a proposition is true if and only if the concept of its predicate is
included in the concept of its subject. Once again, true propositions
are always explicitly or implicitly identical propositions: 'The
predicate or consequent, therefore, is always in the subject or
antecedent, and this constitutes the nature of truth in general … In
identities this connection and inclusion of the predicate in the
subject is explicit; in all other propositions it is implicit and must be
shown through the analysis of notions' (A VI, 4, 1644).

Contingency

When Antoine Arnauld read about these views in a summary of
the *Discourse on Metaphysics* that Leibniz had prepared for him,
he was greatly alarmed by their implications for contingency,
namely, by their implications for the possibility that things *could*
have been different from the way in which they actually are. If
in every individual substance are implied all its predicates, Arnauld
objects, that is, if in the concept of Adam is implied all that he will
do, then everything that happens to Adam will follow from his
complete concept with absolute necessity (A II, 2, 8). If there is not
only *a reason*, but a *sufficient* reason for something to happen, this
something cannot but happen. Are Leibniz's truths of reason
(implying necessity) and truths of fact (meant to underpin
contingency) not so different after all? Leibniz asked himself this
very question: 'If all propositions, including contingent ones,
resolve into identical propositions, are they not all necessary?' (*De
necessitate et contingentia*; A VI, 4, 1449). His answer is a firm 'no'.

In fact, throughout his life, Leibniz strenuously attempted to defend
contingency. A first kind of defence is rooted in the Scholastic

tradition and hinges on the distinction between absolute necessity and hypothetical necessity, mirroring, in turn, the distinction between truths of reason and truths of fact. There is absolute necessity when the negation of a proposition implies a contradiction. For instance, take the proposition: 'a triangle is a three-sided polygon'. Its negation ('a triangle is not a three-sided polygon') implies contradiction because it amounts to saying that a triangle *is not* a triangle. There is hypothetical necessity when something is or happens in a certain way not because it could not have been otherwise, but because it follows from a certain set of preconditions or hypotheses. In the latter case, the opposite does not in itself imply contradiction. For instance, say I live in a shabby one-bedroom flat with no central heating because this is the only place I could afford in London on my own. Given my financial circumstances, I may conclude that it is necessary for me to live in such a dump. This necessity, however, is merely the result of these concrete circumstances. That is, it would not imply contradiction for me *not* to live there, had some conditions been different (had I had more money, or had I married a wealthy man). As Leibniz puts it:

> *Absolute necessity* is when a thing cannot even be understood to be otherwise, but it implies a contradiction in terms, e.g., three times three is ten. *Hypothetical necessity* is when a thing can be understood to be *otherwise* per se, but because of other things already presupposed outside itself, it is necessarily such and such *per accidens*. (A VI, 4, 1377)

Leibniz's second and favoured solution to the problem of contingency emerges around mid-1686 as an 'unexpected light' shining from an unlikely source, 'namely from mathematical considerations on the nature of the infinite' (A VI, 4, 1654). It is recorded in his private papers but not employed either in the *Discourse on Metaphysics* or in the correspondence with Arnauld. This solution hinges on the thesis that, in the case of necessary truths, the inclusion of the predicate in the subject of a proposition is demonstrable, whereas in the case of contingent truths this inclusion is not demonstrable. More precisely, a demonstration implies a finite number of steps, that is, an analysis which comes to an end point. This is possible for necessary truths such as those involving the abstract notions of mathematics. Abstract (or 'incomplete') notions, precisely because they are abstract, involve only a finite number of predicates. Hence the analysis can be conducted to the end.

In the case of contingent truths, however, the analysis is infinite since they involve notions embracing infinitely many predicates, connecting everything with everything. Although the predicate is included in the subject also in contingent truths, it is not possible to demonstrate this inclusion, since their analysis is an infinite process in which the end is never reached. Only God—who (as tradition teaches) does not need the demonstrative reason typical of limited beings but knows everything by intuition—immediately sees the containment of the predicate in the subject in all kinds of propositions, including contingent ones:

> And here is uncovered the hidden difference between necessary and contingent truths, which is not easy to understand if one does not have at least a smattering of mathematics. In necessary propositions, of course, one arrives at an identical equation by means of an analysis continued to a certain point; and this is precisely what it means to demonstrate truth with geometrical rigor. In contingent propositions, however, the analysis proceeds infinitely [in infinitum], through reasons of reasons, so that there is never a full demonstration; nevertheless, the reason of the truth always subsists, although it can be perfectly understood only by God, who alone can go through the infinite series in a single mental apprehension. (*De contingentia*; A VI, 4, 1650)

In short, in the case of both necessary and contingent truths, the predicate is included in the subject, but whereas the demonstration of this inclusion is available to us in the case of necessary truths, such a demonstration is not possible in the case of contingent truths because they involve the infinite. This is why the adequate knowledge of individual substances is open only to the infinite mind of God. Human beings will never be able fully to grasp the complete concept corresponding to any individual substance. As Leibniz writes to Arnauld on 14 July 1686:

> the notion of myself in particular and of every other individual substance is infinitely more extended and more difficult to comprehend than a specific notion like that of sphere, which is only incomplete and does not include all the circumstances necessary in practice to arrive at a particular sphere....As a result, though it is easy enough to judge that the number of feet in the diameter is not included in the notion of the sphere in general, it is not so easy to judge with certainty (thought it can be judged with enough probability) whether the voyage which I plan to make is included in my notion; otherwise it would be as easy to be a prophet as to be a geometer. (GP II, 52–3)

The perfect reason of truths of fact or contingent truths is therefore unknown to us (cf. *Monadology*, § 32), but known to God, because He is the only one able to embrace the infinite.

It seems, however, that Leibniz is not yet off the hook of necessitarianism. If the difference between truths of reason and truths of fact reduces to what we, limited human beings, can know or discover, there is still no genuine contingency. The predicate *is* included in the subject, and all properties or qualities of a certain individual *are* necessarily part of its identity, of what it is to be that individual, whether or not we can discover this inclusion. Although something is or happens in a certain way not because it could not have been otherwise, but because it follows from a certain set of preconditions or hypotheses, if those preconditions (or set of sufficient reasons) *are* given, that something still necessarily follows.

Ultimately, contingency is (narrowly) rescued in Leibniz's system through his doctrine of possible worlds. There are infinitely many possible worlds in which a certain given individual does not exist. Indeed—given the connection of everything with everything and the denial of purely extrinsic denominations—any individual is world-bound, namely it exists in one and only one possible world. It was possible for God to create any one of those infinitely many worlds in which this individual does not exist. Hence it is *not* necessary for this individual to exist because it *could* have not existed. Its existence is the result of a free divine action (A VI, 4, 1449). Hence this individual (and, for that matter, the possible world which was actually created) is contingent.

This conception of contingency, however, turns out to be significantly thinner than even the above rescue operation might, at first glance, suggest. Although it is not *logically* or *metaphysically* necessary for God to create the possible world which, as a matter of fact, he creates, it turns out that, for Leibniz, it is *morally* necessary. It is morally (as opposed to logically or metaphysically) necessary for the wise to choose the best. To use an analogy, although it is not strictly impossible for a loving and wise mother, who is also well aware of what she is doing, intentionally to harm her child gravely and purely for fun (that is, although she has the *power* to do such a thing), this action is morally impossible (as long as she remains loving and wise). God, being supremely good and supremely wise, could not (*morally speaking*) create any world but the best of all possible worlds.

Knowledge

Leibniz's complete-concept theory, and his theory of truth as inclusion of the predicate in the subject, also lead to his theory of knowledge. Knowledge is, basically, seeing (immediately or by a process of discovery) the inclusion of the predicate in the notion of something. If we had a perfect grasp of the complete concept of an individual substance, it would be possible for us to know a priori (that is, independently of experience) what is true of that individual substance by analysing its notion. This analysis is not, however, available to us for most truths. Hence we discover the inclusion of the predicate in the subject a posteriori, on the basis of experience.

Likewise, there are two ways of knowing the possibility of something: a priori and a posteriori. We know the possibility of something a priori when we resolve the concept into its elements or into other concepts the possibility of which is known. If we have carried our analysis to the end and no contradiction has appeared, the possibility of this something is demonstrated. It should be clear, at this point, that such a priori demonstration is possible only for concepts which include a finite number of predicates: if the concept includes infinitely many predicates, there would be no end to the analysis and hence no demonstration. We know the possibility of something a posteriori when we experience the actual existence of the thing, for what actually exists or has existed is in any case possible (A VI, 4, 589–90).

In turn, according to Leibniz, there are different degrees of knowledge depending on the degree to which we can see, or discover, and enumerate the properties or requisites which enter into the notion of something. More precisely, Leibniz distinguishes between knowledge which is *clear but confused* and knowledge which is *clear and distinct*:

> A notion is *obscure* which does not suffice for recognizing the thing represented, as when I merely remember some flower or animal which I have once seen but not well enough to recognize it when it is placed before me and to distinguish it from similar ones … Knowledge is *clear*, therefore, when I have that from which I can recognise the thing represented, and this [clear knowledge] is in turn either confused or distinct. It is *confused* when I cannot enumerate one by one the marks which are sufficient to distinguish the thing from others, although the thing really has those marks and requisites in

which its notion can be resolved: thus we know colours, odours, flavours, and other particular objects of the senses clearly enough and discern them from each other but only by the simple evidence of the senses and not by marks that can be expressed. So we cannot explain to a blind man what red is, nor can we explain such a quality to others except by bringing them into the presence of the thing and making them see, smell, or taste it, or at least by reminding them of some similar perception they have had in the past. (*Meditations on Knowledge, Truth, and Ideas*; A VI, 4, 586)

Distinct concepts are those which enable us to distinguish something from everything else by the enumeration of sufficient marks or requisites, such as when we have a definition. We may, however, 'also have distinct knowledge of an indefinable notion when this notion is *primitive*'. Primitive concepts are concepts which are indefinable because they are 'non-resolvable', that is, they are not composed and can therefore be understood only through themselves (A VI, 4, 587).

When 'all that enters into a distinct notion is in turn known distinctly, that is, when the analysis is carried through to the end, knowledge is *adequate*'. With respect to human cognition, only abstract universal knowledge can be adequate, since only concepts which include a finite number of requisites (such as abstract notions) can be analysed 'to the end'. In most cases, and 'especially in a longer analysis, we do not gaze simultaneously at the entire nature of a thing, but use signs instead of things' (A VI, 4, 587). This last mode of cognition is what Leibniz calls blind or symbolic thought, and it is opposed to intuitive knowledge (A VI, 4, 587–8): 'Of distinct primitive notions there is no other knowledge than intuitive knowledge', but 'of composite notions there is for the most part only symbolic thought' (A VI, 4, 588).

When the analysis of the subject is not complete, our knowledge of the subject is imperfect or inadequate despite including perfect knowledge of some aspects of the subject: 'it is possible to have perfect knowledge [*scientia*] of several truths about objects of which we do not have adequate knowledge [*notitia*]' (A VI, 4, 590). Leibniz is thus able to consider sense-perceptual knowledge as knowledge of its object, although it is confused knowledge. It is important to realize that, for him, most of our knowledge is indeed confused knowledge (or, at best, distinct but not adequate knowledge), and that a primary cognitive goal of human beings is to have the kind or degree of certainty which is sufficient for acting.

Innatism and 'minute perceptions'

Moreover, according to Leibniz,

> although the senses are necessary for all our actual knowledge, they are not sufficient to provide it all, since they never give us anything but instances, that is, particular or singular truths. But however many instances confirm a general truth, they do not suffice to establish its universal necessity; for it does not follow that what has happened will always happen in the same way. (NE 49)

From the fact that the sun has risen thousands of times it does not follow that it will necessarily rise tomorrow. Insofar as we come to know some truths as necessary, this knowledge cannot come solely from experience since, on the basis of experience, we can at best arrive at the generality of a truth, not at its necessity. Leibniz therefore defends the presence in knowledge of an innate element. In his book-length response to Locke's *Essay Concerning Human Understanding*, he writes:

> Our disagreements concern points of some importance. There is the question whether the soul in itself is completely blank like a writing tablet on which nothing has yet been written—a *tabula rasa*—as Aristotle and the author of the *Essay* maintain, and whether everything which is inscribed there comes solely from the senses and experience; or whether the soul inherently contains the principles

of various notions and doctrines, which external objects merely rouse
up on suitable occasions, as I believe and as does Plato. (NE 48)

Experience actualizes or brings to life inclinations, dispositions,
potentialities which are already in the mind, like a match which
needs to be struck in order to catch fire. Thus Leibniz is prepared
to consider the stimulus of experience as necessary for 'all' our
actual knowledge. His point is that experience is not sufficient to
account for all of it. We learn necessary truths on the occasion
of experience but 'we contribute something from our side'
(NE 49), since the demonstrative force of our reasoning is
grounded in innate, active dispositions of our mind to employ
certain principles according to which any sense-data are organized
and evaluated:

> the senses are certainly needed to obtain certain ideas of sensible
> things, and experiences are necessary in order to establish certain
> facts and even useful for verifying reasoning as if by a proof of sorts.
> But the force of demonstrations depends on intelligible notions and
> truths—the only ones capable of making us judge of that which is
> necessary. (GP VI, 503–4)

Hence, according to Leibniz, one should say that 'there is nothing
in the intellect that was not in the senses, except the intellect
itself' (A VI, 2, 393; A VI, 6, 111).

To illustrate these views he uses, amongst others, two similes: a
block of marble in which the veins already mark out the statue of
Hercules (experience acting as a chisel bringing to light the figure
which was already virtually contained in the block of marble),
and a screen ('une toile') which is 'not uniform but is diversified by
folds representing items of innate knowledge' and on which images
(that is, data originating from sense-experience) are projected
(A VI, 6, 52; A VI, 6, 144). It is in this way, Leibniz concludes, 'that
ideas and truths are innate in us, as inclinations, dispositions,
habits, or natural virtualities', and not as actual thoughts.

More generally, according to Leibniz, there is a great deal in our mind of which we are not conscious or aware, namely, infinite 'minute perceptions' which often just amount to some indistinct background noise. This indistinct, confused noise is formed by many discrete sounds, although we are unable to discriminate between them (much the same way in which we are unable to discriminate between the sounds of all the little waves which together result in the roaring of the sea) (A VI, 6, 54). This striking thesis that there are infinite 'minute perceptions' unconsciously crowding our mind is nothing else than the way in which the metaphysical doctrines of the connection of everything with everything, and the mirroring of the entire universe in each individual substance, are captured by Leibniz from the point of view of human cognition. All the way back to the first infancy, which can no longer be consciously retrieved, our mind bears infinite confused traces, signs, echoes of what happens around us—indeed, Leibniz thinks, bears confused traces of what happens in the entire universe. And yet, these 'minute perceptions' of which we are unaware, or only very confusedly aware, may trace the reasons of seemingly inexplicable behaviours, likes, dislikes, and so on. For instance, you may inexplicably hate spaghetti with pesto until your mum tells you that she cooked spaghetti with pesto every time your insufferable great-aunt (of whom you have now luckily lost all memory) came to visit. As Leibniz says, there is a reason for everything, even if most of the time we don't know what it is.

Chapter 6
The best of all possible worlds and Leibniz's theodicy

Of all the infinitely many worlds which are possible, God created (or, more precisely, 'actualized') the best. Leibniz's claim that the actual world, in which we find ourselves, is 'the best of all possible worlds' is deceptively simple and easily caricatured. It was mercilessly satirized by Voltaire (1694–1778) in *Candide* (1759), which recounts the young protagonist's indoctrination at the hands of his pompous tutor, Pangloss. 'Everything is necessarily for the best,' Pangloss explains confidently to his open-hearted pupil. 'Note that noses have been made to support spectacles: so we have spectacles. Legs are obviously designed to wear breeches, and we are supplied with them.'

Leibniz, however, is not a Panglossian who, having looked here and there, comes to the simple-minded, optimistic conclusion that things are really not that bad, despite recurrent pestilence, devastating mass starvation, and the odd shattering earthquake. His claim that this is 'the best of all possible worlds' is based not in superficial observation of the world in which we live, but in the complex logical and metaphysical machinery discussed so far. More precisely, it is an a priori claim in the sense that it does not follow from an observation and evaluation of the balance of good and evil in experience to the conclusion that, all considered, this is the best one can reasonably expect. It follows instead, independently of whatever experience we may have of the quantity and quality of evil mixed with good, from a consideration of the attributes of God, the existence of whom Leibniz regards as a demonstrable truth.

God's existence and the moral necessity of choosing the best

In embracing the demonstrability of God's existence, Leibniz is far from unusual, since this was a view widely shared by early modern philosophers, including Descartes, Locke, and Spinoza. For Leibniz, there are three main ways to demonstrate the existence of God. According to the order in which they are listed in paragraph 45 of the *Monadology*, the first one is a version of the argument already presented by St Anselm (1033–1109). This is an a priori argument since it is based on a consideration of the concept of God, independent of experience. If we stipulate that 'God' means 'the most perfect Being' (or *Ens Perfectissimum*), it follows from the concept of 'most perfect Being' that the 'most perfect Being' must exist. Leibniz points out, however, that before reaching this conclusion one must investigate whether a 'most perfect Being' is possible. Although we think the words 'most perfect Being', it could be that these words do not pick out any possible being, the same way in which there is no possible being corresponding to the words 'square circle'. In God or the *Ens Perfectissimum*, however, there is absolutely no negation, no limitation of any kind. Since 'nothing can prevent the possibility of that which is without any limits, without any negation, and consequently without any contradiction' (GP VI, 614), 'the most perfect Being is possible, because it is nothing other than pure positivity' (A VI, 4, 626). Once the possibility of the most perfect Being has been established, his existence follows.

The second argument presented by Leibniz is a version of the argument 'from eternal truths' of St Augustine (354–430). It is based on the consideration of which kind of reality can be assigned to non-existing possible beings and to what is true of them. As we have already seen, in the framework of Leibniz's philosophy, if there is any reality in the possibility of things, there must be something actual, something really existing, in which this reality is grounded. This something is God as the 'root of possibility', that is to say, as the ultimate foundation of the essence or nature of things which are merely possible, and of necessary or eternal truths: 'if there were no eternal substance, there would be no eternal truths. And from this too God is proven, who is the root of possibility, for his mind is the very realm of ideas or truths' (A VI, 4, 1618).

Finally, God's existence can also be proved a posteriori, that is, following a consideration of our experience of existing entities. The presupposition of this kind of argument is the intelligibility (rationality) of reality. A certain kind of experience would be unintelligible—that is, irrational, contradictory, impossible—if the conditions of possibility of that experience were not given. Everything that we see and experience—Leibniz writes in the *Theodicy*—is contingent, that is to say, it could have not existed. 'Therefore we must seek *the reason of the existence of the World*, which is the whole assemblage of *contingent* things, and we must seek it in the *substance which carries in itself the reason of its existence,* and which is therefore necessary and eternal' (GP VI, 106). In other words, without the existence of a necessary Being there would be no sufficient reason for the existence of the contingent beings of which we have experience (*Monadology*; GP VI, 614). This necessary Being must also be intelligent and endowed with will and power, since intelligence and will are needed to choose a possible world amongst infinitely many, and power is needed to actualize this choice (GP VI, 106).

Now, God being God—namely, the most perfect Being encompassing all unlimited perfections, including absolute power (omnipotence), absolute knowledge (omniscience), and absolute goodness (omnibenevolence)—he cannot but choose the best. An omniscient being can never fail to know what is best. Being also absolutely good, he can never fail to choose it. Being also omnipotent, he cannot fail to bring it about. In brief, God being God, he is morally (although not logically or metaphysically) necessitated to choose the best. As Leibniz writes in the *Theodicy*, 'if we could understand the structure and economy of the Universe, we would find that it is made and governed as the wisest and most virtuous would wish, since God could not fail to do thus. This necessity nevertheless is merely moral' (GP VI, 236).

What counts as 'best'? Metaphysical goodness and human happiness

Of course, this does not yet answer the question of what criteria a possibility must satisfy in order to qualify as the 'best'. One could simply reply that we cannot really know what the criteria are for qualifying as best (at least not on a grand, universal scale), but that the very fact that God has chosen *this* option tells us that it must be the best. Minimally, however, one can attempt to give some content to the notion of 'best' by saying that it must mean something like 'as good as possible, given the circumstances'. The question becomes then what counts as 'good'. Following a long tradition, Leibniz basically thinks of goodness as 'being': insofar as something 'is', that is, insofar as there is some degree of 'being' (something or another, as opposed to not being at all, not having any reality of any kind), this counts as 'good'. Thus, the best possible world is the world in which there is the greatest possible combination of beings, with all their perfections (that is, positive qualities), in all their varieties. In turn, the 'perfection' of a being is for Leibniz its 'degree of reality' or its 'quantity of essence'. Hence, the best possible world is the world containing the greatest 'quantity of essence', or 'the most reality, the most perfection, the most intelligibility' (GP VI, 236).

At this point, one may well think that using the maximization of the quantity and variety of 'being' which is com-possible as chief criterion for determining what is the best of all possible worlds—call it the 'maximization of metaphysical goodness'—can easily turn out to be against what is best for human beings. Does God not care about us? In fact, Leibniz thinks that God does care, and a great deal. As rational creatures endowed with reason and free will, we are the only earthly beings who can enter into society with God, forming a republic of minds of which he is the Monarch. The maximization of the happiness of human beings is thus one of the chief goals of God in his selecting the best of all possible worlds for creation. But it is not the *only* goal, not least because God cares about the other creatures as well. As it happens, however, the maximization of metaphysical goodness does not run against the maximization of human happiness. On the contrary, due to the way in which Leibniz conceives of human happiness, it provides the best objective conditions which can lead to the happiness of rational beings.

Happiness, according to Leibniz, is 'a lasting state of pleasure' (PW 83) or 'delight' which consists in 'feeling harmony' (A VI, 2, 485). Harmony, in turn, is the unity in multiplicity of the greatest possible quantity of essence or com-possible beings. In other words, harmony is an expression of perfection or metaphysical goodness, and happiness is nothing other than the perception of perfection. This perception can be achieved in a confused way through the senses ('the confused perception of some perfection constitutes the pleasure of sense', PW 83), but only rational beings can reach its highest degree, that is, a distinct perception of perfection which can lead to the lasting state of pleasure which constitutes happiness.

In brief, happiness is the lasting pleasure of the mind grasping the intelligible order of reality. The tension between, on the one hand, metaphysical perfection, and on the other hand, the happiness of rational beings as (apparently competing) aims of creation, is resolved in a system in which the rational apprehension of metaphysical perfection is identified with happiness. At the same time, this rational apprehension increases the perfection of minds and, in doing so, increases the total metaphysical perfection of the world—a virtuous circle if ever there was one.

Is a 'best of all possible worlds' possible?

All this being said, one could still wish to get off the boat before the journey even begins by denying that the notion of 'best of all possible worlds' is a coherent one. Assuming (as Leibniz does) that there are infinitely many possible worlds, and that the actual world, as many others, is also itself infinite, how can there be a 'best' one? For there to be infinitely many possible worlds means that no matter how many we think there are, there are more. And for this world, as well as many other worlds, to be infinite means that no matter how great is the number of beings we think there are in each of these infinite worlds, there are more. Leibniz rejects the claim that there can be a number which is the greatest of all numbers, since numbers are infinite. Should he not also reject the claim that there can be the best of all possible worlds, given their infinity?

Leibniz's strategy for repelling this (for his purposes, quite devastating) objection, is to appeal to the principle of sufficient reason. The very fact that there is a world shows that it must be possible for there to be a best of all possible worlds. Otherwise, God would have lacked a sufficient reason for choosing this one over all others, and there would be no world at all (*Theodicy*, GP VI, 232).

Those denying the claim that this is the best of all possible worlds could try, however, a different tack. Surely this actual world would be even better if, say, one extra fawn had escaped death in the Thuringian forest fire of 1016 CE? It is at this point that the full force of Leibniz's net of theses regarding possible worlds, universal harmony, purely extrinsic denominations, identity of indiscernibles, principle of sufficient reason, and so on, comes into play. There is no fact, however seemingly minor, that does not affect the entire universe, since each fact is the result of an infinite chain of reasons which explain why it is thus and not otherwise. Had it been otherwise, the infinite chain of reasons branching off in all directions would have been different. Had there been one extra fawn escaping the fire, it would have been a completely different possible world. Once one grants Leibniz's reasoning about God being morally necessitated to create the best, it follows that that world would not have been the best one, no matter how it may intuitively appear to us from our limited point of view.

Moreover, that complex net of theses also entails that, for Leibniz, each individual is world-bound. In order to get our head around this whole business of possible worlds, it may be helpful for us to think of the worlds in which 'I' am in Paris right now and not stuck in London (as I actually am). However, according to Leibniz, these imaginary counterparts of me enjoying coffee under the Eiffel Tower are unfortunately not me at all but other individuals. Had God actualized one of the worlds in which someone *superficially* similar to me is having a good time in Paris right now, I would simply not have existed (except, of course, as a thought in God's mind). Hence Leibniz berates those complaining that God did not put them in some better circumstances by pointing out that, had they not been in these circumstances, they would not have existed *at all* in the actual world. Not a good outcome for them, one may think. But what about those complaining that it would have been better for them as individuals not to be at all? Since they have to endure such enormous suffering in the world in which they exist, some individual may say, it would have been better for them if God had created another possible world in which they do not exist at all. From the point of view of the balance of good and evil affecting all other creatures, however, it still follows from Leibniz's premises that that different possible world would not have been the best overall, no matter how preferable for those actual-world individuals who wish not to be.

Kinds of evil, human freedom, and the vindication of God's justice

Assuming Leibniz's claim that this is the best of all possible worlds is granted, the framework for much of his theodicy is already in place. He would have shown that all the evils of the actual world are logically necessary for greater goods, although he would still have to answer those objecting that the end of achieving greater goods does not in itself justify the means.

'Theodicy' is a word that Leibniz himself fashioned from the Greek words *Theos*, 'God', and *dikē*, 'justice'. The object of Leibniz's theodicy is the vindication of God's justice, namely, the defence of the justice of God against the charges of injustice brought against him by the existence of so much evil in the world. Defending the justice of God means, for Leibniz, defending also his goodness. Justice is specifically conceived by Leibniz as 'the charity of the wise' (A I, 2, 23). 'Charity' means love, indeed the highest kind of love, a notion well established in the Christian tradition as the chief of the three theological virtues (faith, hope, and charity). When this highest kind of love is guided by knowledge—that is, when it is the 'love of the wise'—it amounts to 'justice' in its highest degree, and constitutes the highest of all virtues.

Since, traditionally, all virtues are conceived as 'habits', justice (being a virtue) is also defined by Leibniz as a habit: the 'habit of loving conformed to wisdom' (PW 83). Wisdom is defined in turn as the 'science of happiness'. Thus wisdom is the way to happiness, or to the lasting state of pleasure which consists in a perception of perfection. Achieving happiness is, finally, the aim and highest good of human beings.

Pitted against the good, there are, in Leibniz's view, three kinds of evil: metaphysical evil, physical evil, and moral evil: '*Metaphysical* evil consists in simple imperfection, *physical* evil in suffering and *moral* evil in sin' (GP VI, 115). Metaphysical evil is a metaphysical necessity and cannot therefore be a ground of complaint against God's justice and goodness. Creatures qua creatures are in fact necessarily limited, and any limitation is a kind of imperfection. To say that something is limited is to say that there is some further degree of whatever property is under consideration that this limited being does not have, or that there is some property which this limited being lacks altogether. In brief, it is to say that this being is not the *Ens Perfectissimum* or God. Since it is the very nature of a creature not to be God, Leibniz notes, regarding God as unjust for not changing this fact would be as absurd as resenting God for not making a circle which is not limited by a circumference (Grua, 365). Natural evil, conceived as a kind of evil which is independent of any moral evil of which human beings may be charged, is subsumed by Leibniz under the category of metaphysical evil because it is, ultimately, the result of creaturely limitation. The classic examples of natural evils, presented also by Leibniz, are natural disasters such as earthquakes and tsunamis, or the naturally occurring lack of a due perfection, such as birth defects. Contrary to other traditional proposals, these natural evils should not be linked, for Leibniz, to moral responsibility.

On the other hand, physical evil or suffering is linked to moral responsibility since it is conceived as the punishment for moral evil or sin. Being the consequence of evil freely done by human beings, physical evil is also no ground for complaint against God's justice. Breaking once again with some traditional views, Leibniz acknowledges that some individuals who suffer as a result of moral evil are themselves truly innocent but regards as sufficient for a vindication of God's justice that any suffering innocently endured in this world will be rewarded in the next, when any imbalance will be redressed. Finally, moral evil is the responsibility of human beings since it is the result of their free choices, that is, of their freedom to choose what *appear* good to them (as opposed to what is truly good).

Leibniz thinks, in fact, that his strict determinism (that is, the view that everything is fully determined by the chain of reasons which explains why it is thus and not otherwise) is compatible with human freedom. Amongst these reasons are free choices of human beings since the causal chain which led to a certain action inclined them to a certain choice but did not necessitate it. Of course, whatever they actually choose to do is eternally fixed as one of the facts which belongs to a possible world which God eternally contemplates in his intellect. But this is not incompatible, according to Leibniz, with the inclusion of free choices in this possible world. In order to be free, an action requires for Leibniz three conditions: spontaneity, that is, the absence of coercion of the agent (the action of someone who cannot but follow the bandits, because she is tied up behind their carriage, is not spontaneous and hence it is not free); intelligence, since intelligence is required to see what appears to be good and hence motivate our action; and contingency (*Theodicy*; GP VI, 288). The latter condition is defined by Leibniz as 'the exclusion of logical or metaphysical necessity' (GP VI, 288), that is, it is enough for the freedom of an act that its opposite does not imply any logical contradiction.

In sum, the only way for God to avoid metaphysical evil would have been not to create at all, but this would have meant that no metaphysical goodness (no being, no reality, no perfection apart from God) would have been actualized either. Provided that any innocent suffering is eventually redressed, physical evil, as a punishment for moral evil, is in itself a manifestation of justice and can also be willed by God as a means to greater goods. Moral evil, on the contrary, is never willed by God as a means to achieve greater goods but merely permitted as a consequence of human free choices.

Embracing a traditional Scholastic distinction between 'antecedent' and 'consequent' divine will, Leibniz maintains that God wills each good and does not will any evil, if we consider

God's will before ('antecedently') any qualifying circumstances (notably, which possible beings are possible together). However, if we consider his will 'consequently' or in the light of qualifying circumstances, God wills the best, that is, he wills the combination of those possible beings which are possible together and which constitute together maximal metaphysical goodness. This entails that God's 'consequent' will rejects some goods and permits some evils.

The justice of God and the justice of men

It is of paramount importance to note, however, that, for Leibniz, the essences (or the natures) of possible beings, and all the eternal truths which express facts about those essences, do not depend on God's will. Essences of possible beings and their combinations in possible worlds are not created by God. God's intellect simply embraces and contemplates them as the full range of what is logically possible and com-possible. As logical possibilities they are what they are independently of any divine will. Their independence from God's *will* does not imply, on the other hand, independence from God, since these logical possibilities ultimately express facts about God's essence. The most fundamental eternal truths are ultimately manifestations of the nature of God. The divine nature is essentially rational. All that is possible is a participation in this divine rationality and is articulated in the divine intellect as an idea in God's mind.

Likewise, the 'essence of the just' (or what counts as just) cannot be a matter of what God *wills*. It is instead an expression or manifestation of what God *is*. As such, it is as independent of subjective views as the truths of mathematics. In a *Meditation on the Common Notion of Justice* written in 1703, Leibniz poses a question that many have been pondering since Plato proposed a version of it in the *Euthyphro*: 'It is agreed that whatever God wills is good and just. But there remains the question whether it is good and just because God wills it or whether God wills it because it is good and just: in other words, whether justice and goodness are arbitrary or whether they belong to the necessary and eternal truths about the nature of things, as do numbers and proportions' (PW 45). Following in Plato's footsteps, Leibniz has no doubt: justice and goodness belong to the necessary and eternal truths about the nature of things. In 1706, attacking the position of Pufendorf, Leibniz explains further:

> Neither the norm of conduct itself, nor the essence of the just, depends on his [God's] free decision, but rather on eternal truths, objects of the divine intellect, which constitute, so to speak, the essence of divinity itself...Justice, indeed, would not be an essential attribute of God, if he himself established justice and law by his free will. And, indeed, justice follows certain rules of equality and of proportion [which are] no less founded in the immutable nature of things, and in the divine ideas, than are the principles of arithmetic and of geometry. So that no one will maintain that justice and goodness originate in the divine will, without at the same time maintaining that truth originates in it as well: an unheard-of paradox...as if the reason that a triangle has three sides, or that two contradictory propositions are incompatible, or that God himself exists, is that God has willed it so. It would follow from this, too, that which some people have imprudently said, that God could with justice condemn an innocent person, since he could make it such that precisely this would constitute justice. (Dutens, IV 280; PW 71–2)

In short, there are objective standards of justice which apply to all beings, including God. In defence of God's justice, it would not do to say that what may be just for God is not just for human beings. Divine justice and human justice differ only in degree not in the nature of what justice is.

Chapter 7
What is ultimately real—unity and activity

With the breadth of Leibniz's vision mapped out in these sketches of his theories on possible worlds and on God's choice of the best of them all, we need to dig deeper to uncover his views about what is ultimately real in this actual world of which we have experience. What are the most fundamental, really existing things? Leibniz follows a long tradition in thinking that the most fundamental entities are individual 'substances'. But what is it to be a 'substance'?

Physics and metaphysics

Leibniz's investigation starts from the world of corporeal, sensible objects which we find around us. Plants, animals, and human beings counted amongst the primary substances of Aristotelian metaphysics, that is, the kind of beings which are fundamental in the universe. They were explained by Aristotle and his followers in terms of two constitutive principles—matter and form. In a broadly Aristotelian framework, matter is a substratum in which first one form, then another form inheres. The intuitive picture is that of bronze which receives a form, being shaped, say, into a vase, and then into a statue, and so on. The form realizes, or makes actual, what is merely potential—that is, actualizes the potential of bronze to become a statue. According to the Aristotelian analysis of change, matter is the ultimate subject of change which form brings about. The form is that which determines which kind of substance or being a certain entity is (e.g. a vase as opposed to a statue). Movement, in turn, is a kind of change. 'Matter' and 'form', 'potentiality' and 'actuality', are therefore some of the fundamental explanatory principles of a broadly Aristotelian metaphysics.

In the evolution of key insights of Aristotelian metaphysics over the centuries, forms were further distinguished into 'substantial forms' and 'accidental forms'. Wood and stones, for instance, are natural bodies which are substances in a primary sense. They are 'this something' (*hoc aliquid*) composed of a 'substantial form' which gives actuality to what in itself is merely potential (and therefore of itself non-existent), namely matter, or, more precisely, 'primary' matter, matter in its primary sense. A house, on the other hand, is an artificial composite which depends for its existence on the existence of these primary substances (wood and stones) which are naturally prior to it. Its unity is not the intrinsic unity provided by a substantial form which actualizes what would otherwise be merely potential and non-existent, but the 'accidental' unity provided by an 'accidental form' which supervenes on pre-existing things with their own actuality.

Building on these broadly Aristotelian metaphysical principles, Aristotelian physics took a 'qualitative' approach to the explanation of natural bodies and their movements. Bodies were explained in terms of their qualitative features and their 'natures', which accounted, in turn, for the movement of a body toward its 'natural' place. Thus, for instance, earth and fire were thought to have, respectively, the qualities of 'gravity' and 'levity'. These qualities explained their natural movements—downwards in the case of earth, upwards in the case of fire.

In the early modern period, however, leading thinkers such as Galileo and Descartes asked themselves anew what the best way is to explain the corporeal world that we encounter in our day-to-day experience. The objects of our sense-experience present themselves as extended. In turn, objects can be distinguished from one another because their extension has a certain magnitude or size, and a certain figure or shape. Finally, to explain our sensible experience, we need to consider not only the extension of objects (as determined by a certain magnitude and shape) but also their movements, what they do in certain conditions, how they behave in specific circumstances.

One of the great insights of Galileo was to realize that, in order to develop a rigorous science of nature, only those aspects of nature which are fully intelligible should be taken into account. These are the aspects of bodies and their movements which can be 'quantified' or 'measured'. In other words, these are the aspects of bodies which can be translated into mathematical terms. On the other hand, Galileo writes in one of his letters on sunspots, it is pointless 'tentar l'essenza', that is, to speculate about essences of which we have no 'intrinsic' knowledge ('notizia intrinseca'). From Galileo's proposal to build a rigorous physics by taking into account only 'quantifiable' aspects of nature—basically, extension and movement—the step was brief to Descartes's contention that extension and movement are the only real features of the corporeal universe. Other features of bodies are 'secondary qualities' which derive from the way in which our senses perceive the world but do not really exist in bodies. A single attribute—extension—defines what it is to be a body. To be a body is simply to be an extended thing (*res extensa*) which can be analysed in mathematical terms. For Descartes, the obscure substantial forms of the Scholastics (that is, broadly speaking, those working in an Aristotelian framework as taught in schools and academies in the later Middle Ages and the early modern period) have no explanatory role to play in the world of bodies, and should be rejected. A new metaphysics of bodies conceived essentially as extended things should provide instead the new roots on which an anti-Aristotelian physics can grow.

The mathematization of nature proved extraordinarily successful and constituted the backbone of the revolutionary scientific advances of the early modern period. In particular, it structured a new, mechanistic physics. The 'moderns' or 'innovators' (as they were called to distinguish them, not always positively, from the adepts of the 'ancient' Aristotelian physics or natural philosophy) embraced 'mechanism', that is, a new philosophy of nature according to which the world is analogous to a large machine, or a very complex clock, the workings of which are to be explained through efficient causes in terms of which elements push, pull, hook, and so on, other elements of the mechanism.

In his own quest for an explanation of the natural world, Leibniz fully embraces the programme of Galilean, mathematically based physics. However, while fully committed to the mathematization and mechanical explanation of natural phenomena, Leibniz argues that the primary, quantifiable features of bodies identified by mechanistic physics—extension, shape, and motion—are insufficient to give reason, in the last instance, of the most fundamental principles which govern the phenomena of the corporeal world. As he declares in a very early text, the *Confessio naturae contra atheistas* of 1668–9, bodies are not self-sufficient. The very analysis of the corporeal world calls for the postulation of an incorporeal principle:

> with the excellent improvement of mathematics and the study of the internal features of things through chemistry and anatomy, it has become apparent that mechanical explanations—reasons from the figure and motion of bodies, as it were—can be given for many of the things which the ancients referred either to the Creator alone or to some sort (I know not what) of incorporeal forms. As a consequence, truly capable men for the first time began to try to save or to explain natural phenomena, or as I should say, that which appears in bodies, without assuming God or taking him into their reasoning. Soon after their attempt had met with some little success, although before they arrived at foundations and principles, they proclaimed, as though congratulating prematurely at their security, that they could find neither God nor the immortality of the soul by natural reason...It

seemed to me fully unworthy for our mind to be blinded in this matter by its very light, that is, by philosophy. I began therefore to attend to the investigation of things myself...Setting aside all prejudices, therefore, and suspending trust in scripture and history, I set my mind to the anatomy of bodies, to see whether that which in bodies appears to the senses can be explained without the supposition of an incorporeal cause. At the beginning I readily admitted that we must agree with contemporary philosophers...that in explaining corporeal phenomena, we must not unnecessarily resort to God or to any other incorporeal thing, form, or quality...but that so far as can be done, everything should be deduced from the nature of body and its primary qualities—magnitude, figure, and motion. But what if I should demonstrate that the origin of these very primary qualities themselves cannot be found in the nature of body? Then indeed, I hope, our naturalists will admit that bodies are not self-sufficient and cannot subsist without an incorporeal principle. (A VI, 1, 489–90)

For Leibniz, physics aims at describing and explaining, in a rigorously mathematical and mechanical way, the 'manifest' world—that is to say, the world as it appears, as it is manifested to us in experience. Its purpose is to predict and master these natural phenomena for the benefit of humankind. In brief, physics proper studies phenomena. Although phenomena are manifestations of substances, substances, or what is ultimately real, are not its proper object. They are instead the object of metaphysics. On the other hand, since physical phenomena are manifestations of what is ultimately real, the most fundamental principles investigated by metaphysics are implicit in physics as well.

Physics, however, is an autonomous enterprise. In order to achieve its purposes, it should not, and need not, rely on metaphysical entities. Its explanations should be purely mechanical but, crucially, purely mechanical explanations do not exhaust what can and should be said about the natural world. A further, metaphysical level of explanation is needed to account fully for the corporeal world. The issue with which Leibniz is grappling would be called, in modern parlance, the issue of 'ontological grounding'. The physical world requires some further 'grounding', that is, it is necessary to postulate some further entity or principle as the condition of this very world as we experience it. Without the postulation of this condition, this world is not intelligible, that is to say, we are not explaining why it is the way it is. In brief, metaphysical entities and principles provide the ultimate grounding of the entities and principles studied by physics.

After the *Confessio naturae*, Leibniz's understanding of mathematics and physics was radically transformed, notably during his formative years in Paris (1672–6) and his sojourn in Italy in 1689–90. This transformation gave him inspiration and new conceptual tools for rethinking the natural world. Yet he never turned his back on the fundamental youthful intuition voiced in the *Confessio naturae*, namely, that an incorporeal principle is needed to account for the corporeal world. In the following years he worked tirelessly toward the best way to conceptualize this incorporeal principle.

Unity

In particular, Leibniz notes that matter, conceived by Descartes as extension, is infinitely divisible. Indeed, according to Leibniz, bodies are actually infinitely divided. The infinite divisibility of extension indicates that bodies conceived, à la Descartes, merely as extended things do not have in themselves the intrinsic principle of unity needed to qualify as 'substances', that is, as the kind of being which is metaphysically primary and irreducible to other more fundamental beings. Even if bodies were not conceived as mere extended things but were still conceived as infinitely divided, far from being substances, they would dissolve like 'sand without lime' (GP III, 500).

Leibniz thus shares the traditional Aristotelian view that being and unity 'turn together' (*ens et unum convertuntur*): nothing can be *a* being without having some kind of unity. To cite another classical formula, 'every entity is one' (*quodlibet ens est unum*). In a letter of April 1687 to Arnauld, Leibniz writes in a similar vein: 'To be brief, I hold as axiomatic this identical proposition which is varied only by the emphasis: namely, that *what is not truly ONE being is not truly one BEING*' (A II, 2, 186). Even a pile of stones or a flock of sheep cannot be *a* pile or *a* flock without having some sort of unity. Their unity, however, falls short of the internal, strict unity which would make them truly 'one being', and which is necessary for an entity to qualify as a substance. A substance must be 'one per se', must have intrinsic unity.

Adopting traditional Aristotelian terminology, Leibniz initially calls 'substantial form' the intrinsic principle of unity of a being, thinking of it as something analogous to the soul or mind in rational beings. How is it that I experience myself as one, persistent, unified being? It is the mind, Leibniz notes, that constitutes me as a being with a unified point of view from which the multiplicity of the surrounding universe is perceived, channelled, organized in one perspective. It is the mind, therefore, that provides the intrinsic unity which makes me an individual substance, and allows me to distinguish myself from all other individual substances. Thought, as experienced first-hand in ourselves, is a prime example of the unification in one perspective of the multiplicity of the universe, the 'unity of the manifold', 'the perception of many things at the same time or one in many' (A VI, 2, 282). Something analogous, Leibniz reasons, must be what constitutes the intrinsic unity of any entity which qualifies as a being properly one, that is, as a substance. As the rational soul or mind was traditionally identified with the substantial form of a human being, so the intrinsic principle of unity (which, analogously to the mind, unifies the manifold into one) is its 'substantial form'.

It is the 'substantial form' which gives whatever reality they have to bodies, providing the intrinsic unity which makes the difference between the mere appearance of being some one entity, and being truly one. As Leibniz writes in 1686 to Antoine Arnauld:

> If the body is a substance, and not a simple phenomenon like the rainbow, nor an entity united by accident or by aggregation like a heap of stones, it cannot consist of extension; and one must necessarily conceive of something there which is called substantial form, and which corresponds in some manner to the soul. (A II, 2, 82)

In the *Discourse on Metaphysics*, written in the same year, Leibniz vindicates the metaphysical explanatory power of the traditional doctrine of substantial forms. Substantial forms are of no use in the explanation of the specific workings of natural phenomena and therefore should not enter into the explanations of physics proper. However, when one is trying to give a metaphysical account of what ultimately grounds these phenomena, the doctrine of substantial forms offers a theoretically powerful way of articulating the intuition that intrinsic unity is a necessary feature of what it is to be a metaphysically primary being, or a substance:

> the ancients as well as many able people accustomed to profound meditations...introduced and maintained the substantial forms which are so decried today. But they are not so distant from the truth nor so ridiculous as the common tribe of our new philosophers imagines. I agree that the consideration of these forms is of no use in the details of physics and must not be used in the explanation of particular phenomena. That is where our Scholastics failed, as did the physicians of the past following their example, believing that they could account for the properties of bodies by mentioning forms and qualities without taking the trouble to examine their manner of operation—as if someone were content to say that a clock has the quality of clockness coming from its form without considering in what all this consists...But this shortcoming and misuse of forms must not make us reject something the knowledge of which is so necessary in metaphysics that, I hold, without it one cannot

properly know the first principles or sufficiently elevate the mind
to the knowledge of incorporeal natures and the wonders of God.
(A VI, 4, 1542–3)

Activity

Unity is one condition of substantiality for Leibniz. Activity is the
other. Arguably, these are not two distinct conditions, but two
different ways of capturing the same condition of substantiality
which can be expressed, overall, in terms of 'activity'. What makes
an entity properly 'one' is its being a source of activity. From very
early on, Leibniz endorses with enthusiasm the traditional,
Scholastic motto that 'actions are proper of supposita' (*actiones
sunt suppositorum*), that is, actions are proper of individual
substances. 'The Scholastics commonly laid down that it is proper
to the Suppositum that it is denominated by action; hence the
Rule: actions are proper of supposita. From which it is clear that
the Suppositum, Substance, and the Being subsisting for itself,
which are the same, are also rightly defined by the Scholastics:
that which has the principle of action within itself' (A VI, 1, 511).

Thought, once again, is a prime example of just such an activity.
Our own activity of thinking gives us a first-hand experience of
what it is to be a source of action which unifies a multitude in one
perspective, or in one unified centre. Adopting Aristotelian
terminology, in numerous texts (e.g. GP VI, 352; GP II, 205–6)
Leibniz calls 'entelechy' (from the Greek *entelecheia*, that which
realizes or makes actual what is potential) the intrinsic principle
of action which characterizes a substance. Sometimes he even
suggests that 'entelechy' is simply a synonym for substance since a
substance is constituted by its being the source of its own internal
actions (GP VI, 609–10).

Moreover, Leibniz maintains that the most appropriate way of
understanding what substantial forms are is to conceive them as
principles of activity, that is, as the 'force', or intrinsic 'power'

which constitutes a substance as a unified centre of activity. With the development of his physics, culminating in the *Dynamics* of 1689–90 where the notion of 'force' takes central stage, 'force' or 'power' increasingly become Leibniz's distinctive way of referring to the intrinsic activity which constitutes a substance. In March 1694, in a short article on the notion of substance, he points to 'force' as the pivotal notion of an account of substance at the interface between physics and metaphysics:

> I will say for the moment that the notion of *forces* or power [notionem *virium* seu virtutis] (which the Germans call *Kraft* and the French *la force*), for the explanation of which I have introduced a special science of *Dynamics*, brings the greatest light to bear on the understanding of the true *notion of substance*. (*On the Emendation of First Philosophy*, GP IV, 469)

In his *New System of the Nature of Substances* of 1695, he explores further the need for physics of a metaphysical grounding, declaring that 'the nature of substantial forms consists in force':

> I had gone far into the country of the scholastics, when mathematics and modern authors drew me out again, while I was still quite young. Their beautiful way of explaining nature mechanically charmed me, and I rightly scorned the method of those who make use only of forms and faculties, from which we learn nothing. But afterwards, having tried to go more deeply into the principles of mechanics themselves in order to explain the laws of nature which are known through experience, I realized that the consideration of mere *extended mass* is insufficient... So it was necessary to recall and, as it were, to rehabilitate *substantial forms*, which are so much decried these days—but in a way which would make them intelligible, and which would separate the use which should be made of them from their previous misuse. I found, then, that the nature of substantial forms consists in force, and that from this there follows something analogous to feeling and desire; and that they must

therefore be understood along the lines of our notion of *souls*. But just as the soul ought not to be used to explain in detail the workings of an animal's body, I decided that similarly these forms must not be used to solve particular problems of nature, although they are necessary for grounding true general principles. Aristotle calls them *first entelechies*. I call them, perhaps more intelligibly, *primitive forces*. (WF 11–12)

Leibniz distinguishes two main kinds of force: primitive forces and derivative forces. The forces studied by physics are those which *derive* from the primitive forces which are the object of metaphysics. In other words, the forces studied by physics are modifications or phenomenal manifestations of the primitive forces which constitute substances. In turn, both primitive forces and derivative forces are distinguished into pairs of 'active forces' and 'passive forces'. Primitive active force, Leibniz explains in his *Specimen dynamicum* (1695), is 'nothing else than the first entelechy' and corresponds to 'the soul or substantial form'. Primitive passive force 'constitutes that which is called primary matter in the Schools, if correctly interpreted' (GM VI, 236–7).

Through his notions of 'primitive active force' and 'primitive passive force', Leibniz reinterprets the Aristotelian distinction between form and matter as a distinction between a primitive principle of activity (the substantial form) and a primitive principle of passivity (primary matter) which are constitutive of any (created or limited) substance. To this pair of primitive forces corresponds a pair of derivative forces: derivative active force and derivative passive force. Active force, according to Leibniz, is the force manifested in motion, which is, in turn, of two kinds: dead force, that is, the force expressed in the possibility of motion (such as, for instance, 'a stone in a sling being held in by a rope'), and living force, that is, the force expressed in actual motion (GM VI, 238). Passive force, on the other hand, is manifested in resistance to motion and accounts for the impenetrability of bodies and their inertia.

The fact that individual beings have an intrinsic principle of activity, or are the source of their own actions, is for Leibniz the single most important way in which his theory of substance is sharply distinguished from competing proposals according to which there is only one substance with genuine causal powers—God. If there were not a force or principle of activity in things themselves, Leibniz argues, we should conclude that it is God who *does* everything. This would amount to saying that God *is* everything and that all things are just modifications of the only one divine substance, as maintained by Leibniz's contemporary and favourite polemical target, Baruch Spinoza (GP IV, 508–9).

In sum, Leibniz declares at various points throughout his life that the adoption of mechanism is not incompatible with elements of Aristotelian metaphysics properly understood. Embracing the traditional insight that a substance must be endowed with intrinsic unity and activity, he devises his own version of this view, pointing at our own mind as the paradigmatic example of what it is to be a substance, namely an entity one per se which is the source of its own internal activity. In March 1690 he writes to Arnauld: 'The body is an aggregate of substances, and, properly speaking, it is not a substance. As a consequence, everywhere in the body there must be indivisible, ingenerable, and incorruptible substances which have something which corresponds to souls; [and it must be] that all these substances have always been and will always be united to organic bodies' (GP II, 135–6).

Chapter 8
Monads

From 1695 onward, Leibniz comes progressively to the view that the best way to capture what it is to be a substance in the strictest sense is through the concept of 'monad'. The word 'monad', Leibniz explains, comes from the Greek *monas* 'which signifies unity, or that which is one' (*Principles of Nature and of Grace*, GP VI, 598). It is clear that with the introduction of his concept of monad Leibniz intends to advance, first of all, his long-standing view that, in order to be a substance, a being must be one per se: what better way to press this point (one can imagine Leibniz thinking) than employing, to signify a substance, a word that just means 'unity'? In the unfinished letter of July 1695 to the Marquis de l'Hôpital in which Leibniz uses for the first time the term monad in his own distinctive sense, monad is defined as 'what is genuinely a real unity' (A III, 4, 451; WF 57). Likewise, in several other texts, the terms 'unities', and 'real unities' are used as synonymous of 'monads'.

Unity and simplicity

Leibniz's monads, however, are increasingly further qualified also as 'simple substances'. In the short pamphlet of 1714 which will eventually be known as the *Monadology* (Figure 8),

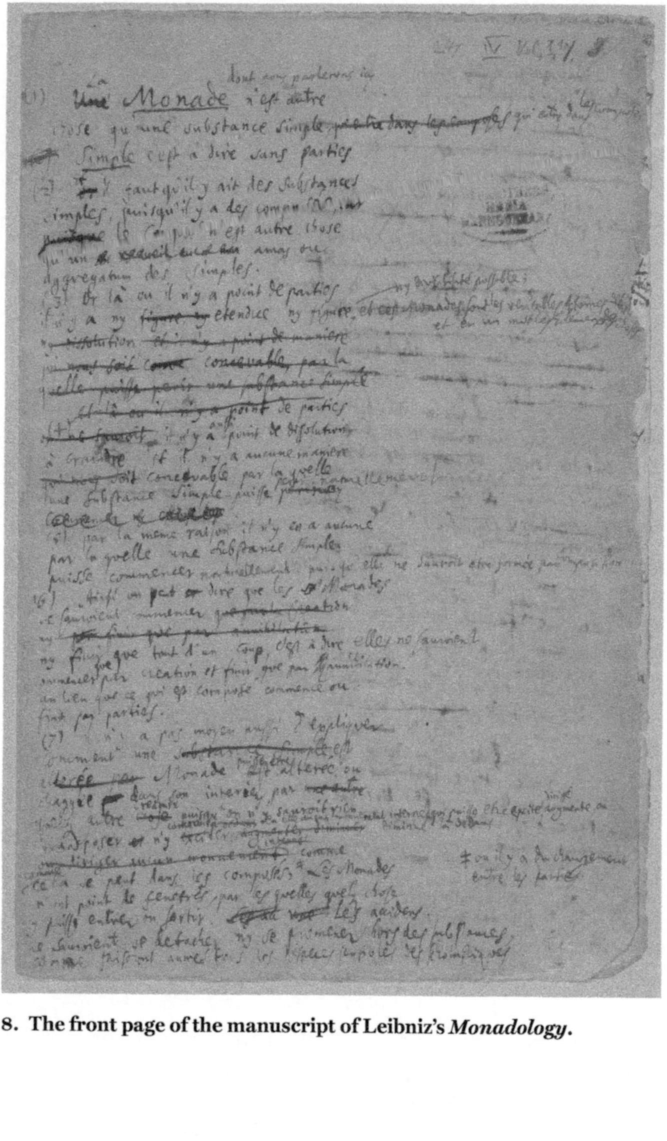

8. The front page of the manuscript of Leibniz's *Monadology*.

Leibniz presents the monad as 'nothing other than a simple substance':

1. The *monad*, of which we shall speak here, is nothing other than a simple substance which enters into composites; *simple*, that is to say, without parts.
2. And there must be simple substances, because there are composites; for the composite is nothing other than a collection or *aggregatum* of simples.
3. Now, where there are no parts, neither extension, nor shape, nor divisibility is possible. And these monads are the true atoms of nature and, in a word, the elements of things. (GP VI, 607)

Why does Leibniz come to think, in the later years of his life, that 'being simple' captures the conditions an entity must meet to qualify as a substance? 'Simple', Leibniz explains, means 'without parts'. A being which does not have parts is indivisible. By Leibniz's lights, indivisibility is crucial for the kind of entities which can be regarded as metaphysically primary or fundamental. Nothing which is further analysable into parts or components can be metaphysically primary because it depends for its existence on these parts or components. These parts or components are the conditions of that which is composed; they are the things which need to exist for the composite to exist. In short, a compound must be ultimately grounded in what is not composed, and lacking composition means being simple. With simple substances we reach what is metaphysically primary, to which other entities can be reduced. In brief, the very fact that there are composites or composed entities (the bodies which we encounter in our day-to-day sense-experience) requires for Leibniz that there must be simple substances.

Moreover, 'simplicity' gives some further content to the broadly traditional claim, endorsed by Leibniz, that substances must have intrinsic unity, must be one per se. This claim raises the question of what counts as per se unity. After all, a flock of sheep, a pile of stones, and an animal body are also one in some sense. When does an entity have a kind of unity which is strong enough to count as per se unity? In his final metaphysics, Leibniz indicates that the strictest kind of unity is the unity enjoyed by 'simple' beings. Since, by definition, they have no parts, they are strictly indivisible and hence 'one' in the strongest way. One may still wonder whether such a strict unity is needed for something to qualify as an entity one per se, endowed with intrinsic unity. This is, in fact, one of the most controversial issues in current interpretations of Leibniz. One thing, however, seems clear enough: whether or not some weaker kind of unity is sufficient, at least 'simple' beings meet the criterion of being one per se, and qualify therefore as substances in the strictest sense.

The simple, indivisible substances which must ground what is composed and divisible are not, however, some indivisible material 'atoms' where the division of matter finally stops. Matter, conceived as extended, is infinitely divisible. There cannot be such a thing as an indivisible extended atom, because the concept of extension entails divisibility. This is precisely the problem Leibniz is grappling with. What can ground the world of extended bodies of which we have experience, if their being extended entails that there is no bottom to their division, no stopping point, and hence no intrinsic unity?

If there cannot be indivisible extended entities which can ground composites, but indivisible entities are a requisite of that which is composed, the only alternative is to postulate indivisible entities which are *not* extended. Insofar as matter is conceived as extended, these entities are also to be conceived as *not* material. Simple, immaterial, non-extended, indivisible entities are the condition of the existence of composed, material, extended, divisible entities. The world of extended bodies studied by physics is ultimately intelligible only if we postulate metaphysical entities which must exist in order for those extended bodies to exist, at least in the minimal sense of being 'something' (whatever that is) as opposed to nothing. Whatever reality extended bodies have, it is borrowed from 'what is genuinely a real unity' (A III, 4, 451; WF 57) and hence 'truly one *being*' (A II, 2, 186). Call them monads or 'unities'.

Activity and force

In the *Monadology*, Leibniz chooses 'unity' as the entry point to his theory of monads. In the *Principles of Nature and of Grace* (a sort of companion pamphlet on monads, also written in 1714), he starts from the other condition of substantiality he has long held to be necessary to qualify as a substance: activity. '*Substance*', Leibniz writes, 'is a being capable of action. It is simple or composite. *Simple substance* is that which has no parts. *Composite [substance]* is the collection of simple substances, or *monads*' (GP VI, 598). As principles of unity, monads constitute the foundation and condition of possibility of multiplicity and composition, which in turn characterize extension. As principles of activity, they constitute the foundation and condition of movement observed in the physical world.

'The substance of things itself consists in the force of acting and being acted upon,' Leibniz writes in an important paper published in 1698, since without an internal force or power in individual things there would be only one substance (God) that *does* everything and hence *is* everything. 'From this one can conclude that there must be found in corporeal substance a *primary entelechy*, or a first subject of activity, namely a primitive motive force ... And it is this substantial principle itself which is called *soul* in living beings, and *substantial form* in other beings, and insofar as it truly constitutes one substance with matter, or a *unum per se*, it makes what I call a monad' (*On Nature Itself*; GP IV, 511). In a letter of 20 June 1703 to the Dutch professor Burchard de Volder, Leibniz explains further: 'I distinguish therefore (1) the primitive Entelechy or Soul, (2) Matter, i.e. primary matter, or primitive passive power, (3) the Monad completed by these two' (GP II, 252).

As outlined in this letter, there are two fundamental aspects to each (created) monad: activity and passivity. They correspond to what Leibniz calls in various texts 'primitive active power' (or 'primitive active force'), 'substantial form', or 'entelechy', on the one hand, and 'primitive passive power' (or 'primitive passive force'), or 'primary matter', on the other hand. Since Leibniz never tires of pointing out that monads are simple, that is, they lack any parts or any composition, their 'primitive active power' and 'primitive passive power' are *not parts* which combine to compose a monad. They are, instead, aspects of a monad which we can think of separately but which are not really separable. Metaphorically and roughly speaking, they are the two sides of the same coin. Technically speaking, they are 'abstractions'—aspects of a simple thing which we analyse in our thought but that do not exist on their own as elements which can be separated from one another.

Any created monad, insofar as it is a substance, must be active, that is, it must be endowed with an intrinsic 'primitive active power'. However, insofar as this monad is a creature (that is, insofar as it is not the *Ens Perfectissimum* or God), this primitive active power is imperfect or limited. In other words, there is in any created monad an element of passivity, labelled by Leibniz 'primitive passive power'. On some occasions, Leibniz uses the Greek word *dynamikon* (dynamism) to refer to a single dynamic principle 'from which [there is] action and passion' (GP II, 241; GP II, 170; GP IV, 394). This single principle of activity and passivity suggests only one real constituent of a simple substance: a *limited* principle of change, which accounts for both the active and passive features of substances, that is, their acting and being acted upon.

In brief, seen from the point of view of activity, monads are unified sources of activity or centres of force. What is ultimately real in the Leibnizian universe is not extended stuff but powers or forces which are manifested derivatively in natural phenomena. Although this idea may have looked highly implausible in the Cartesian world of extended objects impacting mechanically on one another, it no longer seems so odd in the much stranger world of modern physics.

Monads, however, are not centres of forces which can be studied or measured by physics. They are even less the sort of thing that we could see if we had a powerful enough microscope, or any other more advanced scientific instrument. Trying to get our head around monads by imagining tiny building blocks of which extended bodies are ultimately made up, tempting as it is, is not the right approach. Notwithstanding a host of (sometimes quite amusing) metaphors employed by Leibniz, monads cannot be pictured and escape imagination. They belong to a different order, an order accessible by reason and not by the senses. Wishing 'to *imagine* things which can only be *understood*', Leibniz notes, is like wishing 'to hear colours' (GP II, 270). Monads are intelligible entities, that is, entities which we can grasp intellectually as a condition of possibility of bodies but not sense or observe, although observable phenomena are the way in which the primitive forces which constitute monads are manifested in our sense-experience. Leibniz's world of monads which grounds the world of phenomena is a version of the Platonic distinction between reality and appearances, the intelligible world and the sensible world.

Perception and appetite

The properties that Leibniz attributes to monads should make clear that these intelligible entities belong to an order altogether different from the phenomenal world of extended bodies and measurable forces studied by physics. Arguably, we encounter at this point Leibniz's metaphysics at its most audacious. In a suggestive passage of the *New Essays*, he writes:

> When one considers further what belongs to the nature of these real unities, that is *perception* and its consequences, one is transported, so to speak, into another world, that is to say into the *intelligible*

world of substances, whereas previously one was only among the
phenomena of the senses. (NE 378)

All monads, according to Leibniz, are endowed with perception
and appetite. The example which inspires this model is our
own mind and the two fundamental features that we experience
in our own mental life: its cognitive aspect and its conative
aspect. We experience in ourselves a 'perceptual' activity of
various sorts, leading to various types of cognition, as well as a
conative element, that is, a natural tendency, impulse, or
striving which impels us to pass from one action to another,
from one perception to another, from one thought to another,
and so on. One traditional way to capture these two elements in
rational beings (including God) is through the notions of
intellect and will. According to Leibniz, this is the basic model
which applies to all substances, 'since the nature of things is
uniform, and our nature cannot differ infinitely from the
other simple substances of which the whole Universe consists'
(GP II, 270).

All substances are either minds or mind-like entities insofar as
they are all endowed with perception and appetite. We should
not be alarmed, at this point, by the idea that the laptop on
which I am writing is perceiving me, spying on me with little
invisible eyes, as well as moving by its own volition from one
word to another on the screen. That is, we should not be too
quick to anthropomorphize Leibniz's notions of perception and
appetite. Although our mind provides the version of perception
and appetite of which we have experience at first hand, Leibniz's
notions of perception and appetite are much more general.
By perception Leibniz means the 'representation of multitude
in the simple' (GP III, 574) or 'the expression of the many in
one' (GP II, 311). As he writes in the *Monadology*, 'the passing
state which involves and represents a multitude in the unity
or in the simple substance is nothing other than what is called
perception' (GP VI, 608). Thought, in his early years, had

already been defined as the 'unity of the manifold', 'the perception of many things at the same time or one in many' (A VI, 2, 282).

We fail to realize that perception, defined this way, occurs in all monads, Leibniz argues, when we assume that perception must always be conscious. In fact, perception 'must be carefully distinguished from apperception or consciousness' (GP VI, 608). Even our minds are conscious of (or 'apperceive', to use Leibniz's distinctive term) only a tiny fraction of their perceptions. More imperfect monads—indeed, the vast majority of monads—are not conscious of any of their perceptions.

Perception, as conceived by Leibniz at its most basic, bears only minimal similarity to the mental life with which we are familiar. This similarity is restricted to its being a passing state involving a 'representation of multitude in the simple' or 'the expression of the many in one'. Again, we should not be too quick to anthropomorphize this 'representation' by imagining thinking tables and chairs. To use a rough analogy, take one of those tacky souvenirs which turn blue or red depending on the level of humidity. One could say that a plurality of surrounding factors (the level of humidity being in turn the result of other factors) are unified in the state of the souvenir which 'perceives' them. The souvenir 'represents' in a unified way this combination of factors by turning blue.

It is not difficult to see how all sorts of organisms (from a vegetable seed to a bacterium) 'perceive', in this very broad sense, a plurality of factors which are unified in the state of the organism 'representing' that set of factors. Indeed, in a very broad sense, all things 'perceive' their surroundings and 'represent' them through their states. Thus, although there are no thinking tables, a table would 'perceive' a drastic change in gravity or pressure, and would 'represent' it in its state by floating or disintegrating. Although my laptop does not spy on me with invisible eyes, it does 'perceive' its surroundings (including electricity, the pressure of my finger on a key, and so on) and 'represent' them through its state (e.g. through a letter appearing on the screen). These are, of course, all rough analogies which cannot bridge the gap between the *'phenomena of the senses'* and the *'intelligible world of substances'*. But they can give a rough idea of the generality of the notion of 'perception' Leibniz is after.

Monads, at their most basic, are unities in which a plurality of factors are unified in the passing state of the monad which 'perceives' and 'represents' them. 'Since perception is nothing else than the expression of the many in one,' Leibniz writes in 1706, 'it is necessary that all entelechies or monads have perception' (GP II, 311). In the most perfect monads, some of these perceptions are conscious: a mind may be aware of perceiving the sound of the sea even though it is not aware of all the individual noises which result in the sound of which the mind is conscious. And yet, all those individual sounds are unified in the consciously perceived sound. Most of our perceptions, however, are so many, minute, and confused that we are not aware of them: 'Thus, we normally lack awareness of a familiar noise (such as that of a mill close to our dwelling)' (GP II, 311).

'Appetite', in turn, is also to be taken in its most general sense as 'the striving from one perception to another' (GP III, 575), or as 'the action of the internal principle which produces the change or passage from one perception to another' (*Monadology*, GP VI, 609). Appetite is the 'motor' or the principle of change, the tendency to pass from one perception to another (*Principles*, GP VI, 598). Taken in this general sense, all monads must also have appetite to explain the passage from one state to the other.

Although monads are simple substances, their simplicity 'does not prevent a multiplicity of modifications' (*Principles*, GP VI, 598). By 'modifications' is meant here the infinitely many perceptions of each monad. 'We ourselves experience a multiplicity within a simple substance', Leibniz writes, 'when we find that the least thought which we apperceive involves a variety in its object'

(*Monadology*, GP VI, 609). According to Leibniz, each monad mirrors through its perceptions the whole universe, that is, each monad represents the entirety of the same universe from its own, unique perspective (GP III, 636). Leibniz's doctrine of 'minute perceptions', and his complete-concept theory, provide, respectively, the epistemological and the logical versions of the metaphysical point Leibniz is pressing here.

What distinguishes one monad from another is the degree of confusion or distinction of these perceptions. The degree of confusion or distinctness of perceptions is what determines, in turn, the hierarchy of perfection amongst monads. Starting from God—the primitive Monad (GP VII, 502), or 'the primitive Unity, or the first simple substance' (*Monadology*, GP VI, 614), who knows everything perfectly—there are, in descending order of created monads, monads endowed with reason ('minds'), monads endowed merely with sense ('souls'), and monads 'analogous to souls', that is, monads which are endowed with inferior degrees of perception and appetite, or 'bare Monads' (GP VII, 502).

Leibniz pushes his minimalist metaphysical model even further to the point of accounting for the traditional notions of 'matter' and 'form', and for his own distinctive notions of 'primitive passive power' and 'primitive active power', in terms of the degree of confusion and distinction of the perceptions of monads: 'Substances have metaphysical matter or passive power insofar as they express something confusedly; [they have] active [power] insofar as [they express something] distinctly' (A VI, 4, 1504). Or, as he puts it in the *Theodicy*, 'If [an intelligent creature] had only distinct thoughts, it would be a God, its wisdom would be unlimited; this is one of the consequences of my meditations. As soon as there is a mixture of confused thoughts, there are the senses, there is matter' (GP VI, 179).

Moreover, not only does each monad perceive (however confusedly) the whole universe, each monad is like a 'world apart' (GP II, 444;

GP II, 436). The perceptions and appetites of monads are 'intrinsic denominations' (GP VI, 608); that is, they are qualities intrinsic to each monad. In one of his most iconic metaphors, Leibniz declares that 'Monads have no windows, through which something could come in or go out' (*Monadology*, GP VI, 607), meaning that there is no causal interaction between them. Monads do not push, attract, or somehow influence other monads. The sequence of perceptions representing the entire universe unfolds 'spontaneously', that is, without any external influence, in each monad. Each monad is a 'living mirror' of the entire universe—'living' in the sense of being active as opposed to the mere passivity suggested by the 'mirror' metaphor (C 10). Each, therefore, also mirrors every other, but does not interact with any other monad. The analogy of things perceiving their surroundings therefore falls short in this respect (as in others). To use yet another rough analogy, one should rather think of a world of unconnected computers all actively running the same program of virtual reality. They all harmonize in their shared representation of reality without having any influence on one another.

This image, however, immediately suggests the possibility that, far from there being all these nicely harmonizing monads mirroring one another, there is just one created monad (fortunately for me, this would be my own mind) representing a world which does not exist apart from my representations. Reflecting on the scenario of a solitary monad, Leibniz jokes that the entities he is talking about 'are monads not nuns' (Grua, 395). Things, however, are not so clear cut as Leibniz's quick dismissal may suggest. The certainty that the world represented by each monad is not some sort of dream is merely *moral*, as opposed to metaphysical. That is to say, it would be metaphysically possible for a single monad to exist, without making any difference to the phenomena represented by this monad. It does not seem, however, morally possible that God would create only one monad, without all the other monads which constitute the universe represented by the putatively solitary monad.

'There are only monads'

'True substances are only simple substances, or what I call *monads*,' Leibniz writes in 1716, the last year of his life, 'And I believe that there are only monads in nature, the rest being only phenomena that result from them' (Dutens, III, 499). Everything is reducible to simple substances and their modifications (GP VI, 590). Everything results from monads.

The word 'result' is crucial here. Monads are not 'parts' or 'components' of bodies but bodies result from them. Leibniz uses a mathematical analogy: an ideal, mathematical point is not a 'part' of an ideal line but a line 'results' from points. That is, points are 'requisites' of a line; they are the kind of entities which must exist for the line to exist. Thus, monads are 'not parts but requisites of bodies' (GP VII, 503). Likewise, 'truly substantial unities are not parts but the foundations of phenomena' (GP II, 268). 'All in all,' Leibniz writes in 1695, 'everything comes down to these unities, all the rest, or the resultants, being only well-founded phenomena' (WF 46). In short, with his theory of monads, Leibniz reduces to an absolute minimum the 'ingredients' of reality: 'there is nothing in things but simple substances and in them perception and appetite' (GP II, 270).

Finally, since monads have no parts, they cannot naturally perish (*Monadology*, GP VI, 607). This last point matters a great deal to Leibniz, since from very early on he has been firmly committed to the immortality of the soul. The theory of monads gives him at one stroke just such an immortality: on this account, not only the human soul, but all souls and soul-like entities, in brief, all monads, are indestructible (GP VII, 344). What we take to be death is just a state of stupor or unconsciousness similar to the state of all 'bare' monads (*Monadology* § 24).

In conclusion, in order to understand at least part of Leibniz's motivation in developing such an audaciously counter-intuitive metaphysical model, it is important to mention his equally firm early commitment to what philosophers call 'Ockham's razor'. 'Ockham's razor' is associated with the medieval thinker William Ockham (*c.*1285–1347), whose philosophy can be seen as applying in a rigorous way the (Aristotelian) methodological principle according to which 'entities should not be multiplied beyond necessity'. That is, the best explanation is an explanation that accounts for as much as possible in the most economical or parsimonious way. One hypothesis is to be preferred to a competing hypothesis if it explains more or the same with less. Leibniz's theory of monads is one of the most ruthless applications of Ockham's razor in the history of metaphysics. Its goal is to cut away from an account of reality any reducible entities in favour of minimal commitments as to what kind of beings there really are. In his final years Leibniz advances a most radical claim: there are only monads.

Chapter 9
Monads, corporeal substances, and bodies

Having followed Leibniz's thinking about the ultimate constituents
of reality to its metaphysical ground floor, the next natural step is
to explore the relationship between monads and the extended
bodies of the physical world. The issue of how Leibniz conceives
bodies in his mature metaphysics is one of the most discussed in
recent decades. In particular, specialists are debating whether
Leibniz's metaphysical model allows for genuine corporeal
substances. Leibniz writes repeatedly about corporeal substances
throughout his life: the question is how these writings should
be interpreted.

Open debates

Are corporeal substances quasi-Aristotelian substances, irreducible
to monads, and hence the primary substances of a competing
metaphysical model? Leibniz's strong endorsement of the
traditional notion of substantial form as a principle of unity and
activity actualizing matter can be seen as a revival of Aristotelian
substances—substances that are extended unities of matter and
form, and are genuinely one per se. On the other hand, the notions
of substantial form and matter can also be interpreted in a
manner consistent with simple substances or monads, to which
corporeal substances can be reduced. If the latter interpretation is
followed, do these corporeal substances have per se unity, and

hence properly qualify as substances? Or are they said to be substances only in a broad, non-rigorous way because they lack the per se unity required to qualify, strictly speaking, as substances? To complicate matters further, Leibniz writes of (created) monads as always having a body. One can press the case that Leibniz never abandoned the metaphysical model of genuine corporeal substances in favour of ascribing an ultimate reality to soul-like simple substances. According to this interpretation, monads *are* corporeal substances, since (created) monads are always embodied. The fundamental entities of Leibniz's universe, in this view, are not immaterial monads but living corporeal substances with both mental and physical characteristics. In turn, a related interpretation sees bodies as real because they contain substances, that is, principles of activity, which ground their activity.

Part of this complexity arises from the fact that Leibniz experimented with various models and remained remarkably undogmatic in his genuine openness to rethink and reframe his position in response to objections. Yet the fundamental question still stands: does he have a considered and basically settled final view which can be traced and pieced together through varying formulations and from the different angles from which it is stated? Or is he still working away until his death at a final considered view rather than having achieved one? And, if there are, in his thought, competing metaphysical models, are these different considered positions at different stages of his philosophical development (notably, during his 'middle years', before he started to write of 'monads' and 'simple' substances)? Or did he hold competing models at the same time? If the latter, are these models irreconcilable? If so, was Leibniz fully aware of this? Or are they not irreconcilable, at least by Leibniz's lights? No consensus has emerged in the literature, although some interpretations have gathered broader support amongst Leibniz scholars than others.

The problem is again compounded by the fact that Leibniz never tells the whole story in a single text. There is no *Summa* of his philosophy, or even of his metaphysics. He often develops and clarifies his views in correspondence with friends and fellow savants, under the pressure and stimulation of their objections, Furthermore, he was well aware of the difficulty of getting his full, counter-intuitive metaphysical package across. His strategy was rather to proceed step by step, speaking to his conversation partners within a philosophical framework with which they would sympathize, taking them with him as far as possible on their own terms. Sophisticated cases based on textual evidence have been made for all the interpretations briefly mentioned above. The interpretation provided here gives preponderance to Leibniz's variously formulated claim in his final years that the created world consists, ultimately, only of monads.

A fivefold scheme of things

In the letter of 20 June 1703 to De Volder in which Leibniz presents the monad as 'completed' by a primitive entelechy (or primitive active power) and primary matter (or primitive passive power), his full account of things follows a fivefold scheme:

> I distinguish therefore (1) the primitive Entelechy or Soul, (2) Matter, i.e. primary matter, or primitive passive power, (3) the Monad completed by these two, (4) the Mass [*Massa*] or secondary matter, or organic machine, for which countless subordinate Monads come together, (5) the Animal or corporeal substance, which is made One by the Monad dominating the Machine. (GP II, 252)

Two kinds of matter are distinguished in this passage: primary matter and secondary matter. Primary matter is the aspect of passivity which is proper to all created monads due to the fact that creatures qua creatures cannot but be limited. In other words, the primitive activity which constitutes a created monad is limited; therefore, a created monad is to some extent passive. Only God, or the *Ens Perfectissimum*, does not have any primary matter since God is purely active.

Secondary matter is an aggregate of monads. This is the body, or 'organic machine', of a corporeal substance or 'animal'. What unifies the animal into one entity—that is, what makes it a corporeal substance endowed with some kind of unity, as opposed to being a mere aggregate of monads—is a monad which 'dominates' the machine. In embodied human beings, the 'dominant monad' which provides the principle of unity is the mind. The 'machine' is the body, ultimately resulting from an aggregate of monads. The same model, according to Leibniz, applies to all created entities: a non-rational animal is also made one by a dominant monad (the soul) which provides the animal's principle of unity by dominating its body, 'or organic machine, for which countless subordinate Monads come together'; and all other bodies are also ultimately aggregates of monads dominated by a monad. In brief, the general account of any corporeal substance is as follows: a corporeal substance is constituted by an infinite aggregate of monads (secondary matter), unified by a dominant monad fulfilling the role of substantial form.

Each monad of the aggregate which constitutes the organic body of the dominant monad, however, is in turn the dominant monad of an organic body, and so on to infinity. Each created single monad, therefore, always has a body (constituted by other monads). As Leibniz summarizes in the *Principles of Nature and of Grace*, 'each distinct simple substance or monad, which forms the centre of a composite substance (for example, of an animal) and the principle of its oneness, is surrounded by a *mass* composed of an infinity of other monads, which constitute the body belonging to this central monad' (GP VI, 598–9). In contrast to monads, a corporeal substance is a 'composite substance', or a 'compound', formed by simple substances or monads (*Principles*, GP VI, 598; *Monadology*, GP VI, 607). Although each corporeal substance is constituted by an infinite envelopment of other corporeal substances, ultimately everything reduces to monads.

At this point, the question to be addressed is how this unification takes place, given the uniform nature of monads as simple substances endowed with perception and appetite. In a letter of 16 June 1712 to his friend the Jesuit Bartholomew Des Bosses, Leibniz explains that 'domination and subordination of monads, considered in monads themselves, consists only in degrees of perception' (GP II, 451). Here the effects of Ockham's razor are again in full view: once everything has been reduced to simple substances and their internal states of perception and appetite, the constitution of corporeal substances as entities endowed with some kind of unity is also explained purely in terms of monads and their perceptions. The organic body of a dominant monad is constituted by those monads which are perceived more distinctly by the dominant monad. Given that there is no causal interaction between monads, this more distinct perception appears to mean that the dominant monad mirrors the states of its subordinate monads in a more distinct way than the states of other monads which are not perceived as constituents of its body, the states of which are mirrored or represented only in a very confused way. This is, basically, the way in which union is explained in any corporeal substance, including an embodied human being.

Pre-established harmony, and the union of mind and body

Leibniz labels this model of union 'pre-established harmony'. Although this harmony applies to all corporeal substances, its full development is driven by the sustained debate besetting early modern philosophy regarding the nature of the union between mind (or soul) and body in a human being. In the framework of Descartes's philosophy, the mind–body problem had become particularly acute. According to Descartes, there are two kinds of created substance: mind, that is, the *res cogitans* or 'thinking thing'; and body, that is, the *res extensa* or 'extended thing'. The nature of these two kinds of substance is utterly different: one is immaterial and consists in thought; the other is material and consists in extension. This dualism of sharply distinct natures immediately raises the question of how something immaterial can interact with something material and form a single being. Some offshoots of Cartesianism proposed a metaphysical model known as 'occasionalism'. According to this model, God coordinates what happens in the body with what happens in the mind, with no causal interaction between mind and body. Thus, on the 'occasion' of a bodily motion, God will cause an appropriately corresponding thought in the mind, and so on.

Leibniz's model of the union of mind and body bears some superficial similarities to occasionalism. Below the surface, however, the differences are deep, as Leibniz himself was keen to stress. According to Leibniz, there is no ad hoc divine coordination between what happens in the body and what happens in the mind. Rather, God has 'pre-established' a complete harmony between the two. This harmony is eternally inscribed in the very nature of the mind and body which are united, and which will then unfold in perfect synchronicity without any direct influence on one another or any need for God's intervention. What the occasionalists explain with continuous miracles, Leibniz proudly claims, his system explains naturally: the system of pre-established harmony explains the union of mind and body in terms of the agreement and harmonious unfolding of the nature of substances.

To understand why Leibniz was so thrilled by his system and regarded it as profoundly different from occasionalism, it is important to realize how this model dovetails with different strands of his thought. If we remain at the stage of logical possibility and consider the complete concepts of possible individuals which combine into a possible world by virtue of their com-possibility, all these complete concepts already include everything that will happen in that possible world. Likewise, if we consider the actual, created world of simple substances each mirroring the whole universe, these simple substances have natures in which all that will happen to them and their world is already inscribed. These natures will unfold spontaneously (that is, without external influences) according to plan, all in the strictly coordinated way dictated by their com-possibility.

The key difference between Leibniz's system of pre-established harmony and occasionalism is precisely this *spontaneous* unfolding or 'spontaneous internal changes' (GP II, 276), allowing for intra-causation between unfolding internal states, but not for inter-causation between distinct beings. The fundamental entities of the system have genuine, intrinsic causal powers which constitute them as unified active beings with intrinsic principles of unity and activity. In a word, their intrinsic causal powers constitute them as *substances*. In the occasionalist system, as Leibniz sees it, due to the lack of such spontaneous unfolding, the only genuine causal powers are God's. Despite all the good intentions of its followers, Leibniz therefore regards occasionalism as heading down the slippery slope of Spinozism, in which there is only one substance and all things are merely its modifications.

Finally, the body involved in the pre-established harmony model of union is not a 'Cartesian' body, that is, a *res extensa* with a nature sharply different from an immaterial *res cogitans*. It is, instead, a 'Leibnizian' body, namely a body which ultimately reduces to an aggregate of monads perceived more distinctly by the dominant monad. The pre-established harmony of mind and body is therefore merely a special case of the universal, pre-established harmony amongst all monads.

Corporeal substances and unity

Leibniz was therefore not without justification in regarding his metaphysics, however anti-intuitive, as rigorous, elegant, and beautifully simple in principle. One can consequently appreciate his annoyance when sharp-eyed readers of his *New System of the Nature of Substances and their Communication, and of the Union which Exists between the Soul and the Body*, published in 1695, objected to him that pre-established harmony is not, after all, very different from occasionalism insofar as it also does not account for a true union between mind and body. Mind–body agreement and harmony is all very well, but still falls short of real, metaphysical union.

The root of the problem is whether Leibniz's beautifully economical theory of simple substances out of which everything else results has the resources to grant the status of genuine substances to classical Aristotelian primary substances, such as plants, animals, and embodied human beings. In a monadological framework, these are all beings composed of substances unified by a dominant monad. Is this unity strong enough to count as per se unity of the unified entity? Can monadic domination account for an intrinsic unity?

According to Leibniz, any aggregate is mind-dependent. In order to have an aggregate, there must be a mind thinking together (that is, unifying) the elements of which the aggregate is made. As he writes to De Volder: 'an aggregate is in fact nothing other than all the things from which it results taken together, which certainly have their Unity only from a mind, on account of those features which they have in common, like a flock of sheep' (GP II, 256). The case of substances unified by a dominant monad into a corporeal substance seems to be different from the case of a flock of sheep in one important respect. Unlike the flock, in which the unification of different elements into one entity is due to a mind external to the aggregate, the mind or mind-like entity which unifies 'countless monads' into a corporeal substance by perceiving them more distinctly is 'internal' to the composite itself. One could press the point that this counts as an intrinsic principle of unity, hence allowing corporeal substances to be regarded as genuine substances which meet the criterion of being one per se.

The problem is, however, that the kind of unity we are confronted with is still a merely relational unity, that is, a unity purely based on how distinctly the dominant monad represents other monads with which it has no direct causal interactions. It seems at least doubtful that such unity, although internal to the composite, can count as the intrinsic, per se unity, required for substantiality. As a relational unity, the unity of a corporeal substance is still an ideal unity, as opposed to a real one. Despite being grounded in the internal states of monads which a dominant monad perceives more distinctly, it still amounts to an agreement amongst substances which perceive or express each other, but have no real connection with one another. The unity of a corporeal substance is therefore of a stronger kind than the unity, say, of a flock of sheep or a pile of stones, but it is not at all clear that Leibniz can claim for it the per se unity constitutive of a genuine substance.

In his correspondence with Des Bosses, Leibniz confronts the problem head-on. From 1712 onward his letters explore the possibility of a metaphysical union superadded to monads and providing the ultimate principle of unity of a composite substance. After various experiments, he candidly confesses, however, that the introduction of such a *vinculum substantiale* or substantial bond over and above monads is not really consistent with his monadological framework. Although Leibniz does not deny the possibility of a stronger principle of unity bestowing genuine substantiality on composite beings, he comes to the conclusion that pre-established harmony and monadic domination will have to do, since they are sufficient to explain the relationship which we perceive between mind and body.

In sum, it seems Leibniz declines to postulate a stronger metaphysical union between mind and body as much for methodological reasons as for any other. Such a postulation should be abandoned not because he has firm views about its existence or non-existence, but because it is unnecessary for the explanation of phenomena which are sufficiently accounted for, in an economical way, within a monadological framework with no room for superadded substantial bonds. Although created monads, for Leibniz, always have a body, he remains remarkably unconcerned about whether his metaphysical model allows the union between dominating monads and their bodies to qualify as the true unity of a genuine substance. In particular, in the case of a human being, Leibniz seems to be quite content with the thought that the mind or soul is indeed a substance, and there is no pressing need to account for a substantial or metaphysical union with the changing body or aggregate of monads which always accompanies it.

Extended bodies as well-founded phenomena

All in all, Leibniz concludes at the end of his protracted exploration with Des Bosses, nothing else is needed apart from monads and their relations of domination and subordination to explain what is manifest in our experience. One may well be taken aback by this conclusion. What about bodies as we experience them? Surely, if anything counts as being manifest in experience, this is extended, resistant bodies, rather than mind-like simple substances.

Yes, of course, we can imagine Leibniz replying, these extended, resistant bodies of which we have experience are indeed the *phenomena* which are manifest to us. That is, what we normally regard as real things are well-founded phenomena resulting from monads. In particular, extension or extended matter is for Leibniz a well-founded phenomenon manifesting something which is not really extended. 'Well-founded' here means that these phenomena are grounded in something which is ultimately real. As its expression, they are not illusions.

Leibniz's favourite analogy is the rainbow. 'Accurately speaking', he writes on 30 June 1704 to De Volder, 'matter is not composed of constitutive unities but results from them, since matter, or more accurately, extended mass, is nothing but a phenomenon founded in things, like the rainbow or the perihelion; and the reality in all things is only that of the unities ... Truly substantial unities are not parts but the foundations of phenomena' (GP II, 268). As the coloured bands of the rainbow are the way in which light refracted through a multitude of droplets appears to us, so extended matter is the way in which an aggregate of non-extended simple substances appears to us. Like any analogy, however, the rainbow can take us only so far in understanding the relationship between extended matter and the non-extended substances from which matter results. The coloured bands and the light-refracting droplets of water which ground them, however different, are in the same physical order of things. Extended bodies and non-extended substances belong to two different orders altogether: the physical order and the metaphysical order, or what Leibniz contrasts in the *New Essays* as the sensible world of phenomena versus 'the intelligible world of substances'.

The details of how physical phenomena may result from an intelligible world of substances are somewhat obscure and perhaps not fully worked out; but Leibniz gives some account of the relationship between the physical and metaphysical order in terms of derivative and primitive forces. The active and passive forces of bodies studied by physics are conceived as expressions of the primitive forces of monads, with primitive active force being manifested in motion, and primitive passive force being manifested in resistance to motion and impenetrability. Thus, the moving forces of physical bodies are the manifestation of the entelechy or active aspect of monads, while extension is explained as the confused manifestation of the diffusion or repetition of natures which have an aspect of passivity (what Leibniz calls the 'primary matter' of monads). As Leibniz writes to De Volder,

> [extension] expresses nothing other than a certain not successive (like duration) but simultaneous diffusion or repetition of the same nature, or what comes to the same thing, a multitude of things of the same nature, existing simultaneously with some order among them...Moreover, this nature that is supposed to be diffused, repeated, or continued is that which constitutes physical body, and

it cannot be found in anything other than the principle of acting and being acted upon, since nothing else is suggested to us by the phenomena. (GP II, 269)

To summarize, there are three clearly distinct senses of matter for Leibniz: primary matter, secondary matter, and extended matter. By 'primary matter' is meant the primitive aspect of passivity of a (created) simple substance or monad. Secondary matter is an aggregate of monads which constitutes the body of a dominant monad. Extended (or 'Cartesian') matter is a phenomenon expressing the way in which aggregates of monads appear to us. The latter is the kind of matter which is the object of physics. There is, therefore, a sense in which matter and bodies are 'real' for Leibniz in a robust, metaphysical manner—the same robust metaphysical manner in which monads are what is ultimately real. Secondary matter and the bodies constituted by it are, ultimately, nothing else but monads. As such, they are as real as anything can be in Leibniz's final metaphysical system. Whether or not such aggregates of monads also qualify as corporeal 'substances' in a strict sense, they are in any case constituted by entities which are without doubt substances.

The reality of bodies resolves, however, into the reality of monads and into the way in which the distinct perceptions of a dominant monad aggregate other monads. Furthermore, whatever reality extended matter and extended bodies have, it is borrowed from the simple substances or unities from which they result, and of which they are the phenomenal expression. Thus Leibniz claims that 'matter and motion are not so much substances or things as phenomena of perceivers, the reality of which is situated in the harmony of perceivers with themselves (at different times) and with other perceivers' (GP II, 270). Or, as he writes in 1714, 'material things and their motions are merely phenomena. Their reality lies only in the agreement of the appearances of Monads.' (GP III, 567n.)

All things considered, is Leibniz doing away with bodies altogether, as his correspondent De Volder fears? No, Leibniz firmly replies:

> I do not really take away body, but bring it back to what it is. For I show that a corporeal mass which is believed to have something other than simple substances is not a substance but a phenomenon resulting from simple substances, which alone have unity and absolute reality. I relegate derivative forces to the phenomena, but I think it is plain that primitive forces can be nothing other than the internal strivings of simple substances...Anything beyond this in things is posited to no purpose and built in addition without argument. (GP II, 275)

Independently of what one may think of the merits of this attempt to give reason of our day-to-day encounters with the sensible, extended world, Leibniz's monadology is not the philosophical fairy-tale of a thinker not really interested in bodies. On the contrary, it is driven by a search for the conditions of possibility of this very physical world of which we have experience. This search comes to the conclusion that the extended world presupposes as its condition non-extended unities in which the phenomena of our sense-experience are grounded.

Epilogue

Leibniz's philosophy is the product of a multi-layered conversation
with centuries of past thought and hundreds of individuals of his
own age scattered all over Europe and as far afield as China. It is
possible to trace important elements of his way of thinking back to
many different traditions, all of them reshaped and remodelled
into a strikingly original outlook. Leibniz himself recommended to
search for and to treasure fragments of truth everywhere. 'Truth is
more widespread than people think,' he noted in a letter to Nicolas
Rémond of August 1714, 'but it is very often disguised and very
often also enveloped and even weakened, mutilated, corrupted by
additions which spoil it or render it less useful' (GP III, 624–5).
His intellectual policy was to stress agreement over disagreement.
'Shame on those who maintain schism through their obstinacy,'
Leibniz wrote in the context of the religious controversies plaguing
his time, 'not wanting to attend to reason, yet wanting to have it
always' (A I, 6, 121). Speaking to the Scholastics, he adapted himself
'somewhat to the language of the School'; engaging with followers
of Descartes of all stripes, he adopted 'the style of the Cartesians';
and speaking to 'those who are not yet very accustomed to either of
these languages', he tried to express himself in a way 'which could
be understood' (GP III, 624–5).

Together with other traditions, Leibniz inherited some fundamental
Aristotelian insights, mediated by their re-elaboration in various

strands of Scholasticism. To the Cartesian universe, divided into two radically different kinds of substances—the world of extended substances (the essence of which is only extension) and the world of thinking substances (the essence of which is only thought)—Leibniz opposed a unified universe in which all substances have a basically uniform nature, and participate in some degree of unity and activity. There is at least one important respect in which this universe is closer to an Aristotelian outlook than a Cartesian outlook, namely in its being—as the Aristotelian universe in which primary substances are unities of matter and form—a metaphysically unified one, with no dualism between two sharply distinguished kinds of substance. On the other hand, Leibniz's thought is deeply marked by views put forward by his contemporaries. Even if Hobbes, Descartes, Spinoza, and Locke (amongst others) constituted some of Leibniz's favourite polemical targets, there are aspects of their world-views which resonate profoundly with Leibniz, and had a significant impact on the shaping of his own thought.

Another key element of this shaping is the 'minimalism' recommended by Ockham's razor as a methodological guiding principle. This is fully visible in Leibniz's austere reduction of all really existing entities to (fundamentally) the same kind of beings, namely unities of activity, in which activity is uniformly conceived as perception and appetite. His minimalism regarding really existing beings is also evident in his conception of essences or natures, mathematical and logical entities, truths, and relations, as purely mental entities (or *entia rationis*): they are not 'things' but 'thoughts'. Leibniz's nominalism, or more precisely 'conceptualism', according to which these entities exist only as 'concepts' in some mind, is wedded, on the other hand, to a broadly Platonic, realist element according to which essences and truths have a reality 'beyond our intelligence' because they are eternal thoughts in the mind of God. As ideas *in mente Dei*, they have objective, immutable natures, independent not only from us but also from God's will.

As Leibniz himself acknowledged, the overall mould, or overall inspiration of his thought, is in fact broadly Platonic. This is a theologized Platonism which reached the 17th century through the filter of Neoplatonism, Augustine, and a host of other interpreters and mediators. This long and multiform tradition places the Platonic intelligible world of forms or ideas in the mind of God. It is God's intellect which grounds whatever reality there is in possibility, and God's nature which grounds the objective nature of the most fundamental principles and moral values. The world which really exists is the world which God actualizes because it is the best amongst infinitely many possible worlds embraced and surveyed by his intellect. In this actual world, things, as they really are, are grasped not by the senses, but by reason. The sensible world of appearances, or the phenomena of the senses, are the expression or manifestation of an intelligible world of substances which constitutes what is ultimately real.

It was in order to explain the actual world as we experience it, and what good and evil we find in it, that Leibniz took us on a journey through possible worlds and the mind of God. This theoretical understanding was for him at the core of an ultimately practical project of scientific advancement for the benefit of humankind. The claim that this is the best of all possible worlds did not mean for Leibniz that it makes no sense to work for its improvement. On the contrary, its being the best implies an infinite progress in which we have a key part to play. As captured in Leibniz's motto for his academies of sciences—*Theoria cum Praxi*—thinking must be wedded to acting. In a number of programmatic statements interspersed throughout the sprawling corpus of his writings, Leibniz traced a remarkably coherent picture of how all his enquiries fit together into an overarching intellectual vision of the systematic advancement of all the sciences. As he announced in his first outline of this encompassing plan in 1668-9, the development of the 'elements of philosophy'—namely the first principles of metaphysics, of logic, of mathematics, of physics, and of ethics and politics or 'practical philosophy'—was needed as

9. *Leibnizhaus* in Schmiedestraße, Hanover. From 1698 until his death on 14 November 1716, Leibniz lived in an apartment of the tallest building in the picture.

prolegomena to further enquiries. Clearly, he did not intend to leave this foundational theoretical work merely to others. Instead, he applied his mind to it throughout his life, while relentlessly proposing and pursuing a flurry of political, social, economic, administrative, and technical schemes and reforms. It is in this way, Leibniz thought, that we celebrate the glory of God and advance the public good: 'by means of useful works and beautiful discoveries'. With his extraordinary intellectual creativity, he contributed to this all-embracing project more than most.

Timeline

1 July 1646	Gottfried Wilhelm Leibniz is born in Leipzig (Saxony).
1661–6	Studies at the University of Leipzig and, for the summer semester of 1663, in Jena. Bachelor and Master degrees in Philosophy, and Bachelor degree in Law.
1666	Enrolment in the Law faculty of the University of Altdorf (near Nuremberg), where he gains Licence and Doctorate in Law.
Autumn of 1667	Leibniz leaves Nuremberg, intending to undertake a European grand tour.
End of 1667–8	First encounters with Baron Johann Christian von Boineburg and employment by the elector and prince-archbishop of Mainz, Johann Philipp von Schönborn.
19 March 1672	Departure for Paris.
End January–end February 1673	First visit to London.
October 1675	Invention of the infinitesimal calculus.
4 October 1676	Leibniz leaves Paris.
18–29 October 1676	Second visit to London.
November 1676	Leibniz in Holland where he meets Baruch Spinoza.

Mid-December 1676	Arrival in Hanover to assume the duties of court counsellor and librarian to Duke Johann Friedrich.
28 December 1679	Death of Duke Johann Friedrich. The duchy of Hanover passes to Ernst August, married to Sophie von der Pfalz.
1680–6	Heavy involvement in the Harz mines.
June 1685	Leibniz tasked with writing the Guelf history.
End of October 1687	Departure for southern Germany.
8 May 1688–February 1689	Leibniz in Vienna.
11 February 1689	Departure for Italy, where Leibniz visits a number of cities (including Venice, Modena, Bologna, Florence, and Naples), and spends an extended period in Rome.
March 1690	Leibniz leaves Venice for Vienna.
Mid-June 1690	Leibniz returns to Hanover.
2 February 1698	Death of Ernst August. His oldest son, Georg Ludwig, becomes the new elector and duke of Hanover.
November 1698–February 1705	Leibniz spends extended periods of time in Berlin (some twenty-four months in total), establishing an exceptionally close bond with the sister of Georg Ludwig, Sophie Charlotte, electress of Brandenburg and (from 1701) queen of Prussia.
March 1700	Establishment of the Berlin Society of Sciences under Leibniz's presidency.
Mid-December 1712–early September 1714	Leibniz in Vienna.
August 1714	Death of Queen Anne of Great Britain and Ireland; passage of the throne to Georg Ludwig of Hanover (George I).
3 September 1714	Departure from Vienna.
14 September 1714	Leibniz arrives in Hanover, three days after Georg Ludwig and his court have left for London.
6 November 1716	Leibniz falls ill.
Evening of 14 November 1716	Death of Leibniz in Hanover.

Abbreviations

A	G. W. Leibniz, *Sämtliche Schriften und Briefe*. Ed. Academy of Sciences of Berlin. Series I–VIII. Darmstadt, Leipzig, and Berlin, 1923ff.
I	*Allgemeiner, politischer und historischer Briefwechsel*
II	*Philosophischer Briefwechsel*
III	*Mathematischer, naturwissenschaftlicher und technischer Briefwechsel*
IV	*Politische Schriften*
V	*Historische und sprachwissenschaftliche Schriften*
VI	*Philosophische Schriften*
VII	*Mathematische Schriften*
VIII	*Naturwissenschaftliche, medizinische und technische Schriften*. Cited by series, volume, and page.
C	G. W. Leibniz, *Opuscules et fragments inédits*. Ed. Louis Couturat. Paris: F. Alcan 1903. Cited by page.
DM	*Discourse on Metaphysics*
Doebner	R. Doebner (ed.), 'Leibnizens Briefwechsel mit dem Minister von Bernstorff und andere Leibniz betreffende Briefe und Attenstücke aus den Jahren 1705–1716', *Zeitschrift des historischen Vereins für Niedersachsen* (1881): 205–380. Cited by page.
Dutens	G. W. Leibniz, *Opera omnia, nunc primum collecta, in classes distributa, praefationibus et indicibus exornata*. Ed. L. Dutens. 6 vols. Geneva: De Tournes, 1768. Cited by volume and page.

GB	*Der Briefwechsel von Gottfried Wilhelm Leibniz mit Mathematikern*. Ed. C. I. Gerhardt. Berlin: Mayer & Müller, 1899. Cited by page.
GM	G. W. Leibniz, *Mathematische Schriften*. 7 vols. Ed. C. I. Gerhardt. Berlin and Halle: A. Asher and H. W. Schmidt, 1849–63. Cited by volume and page.
GP	G. W. Leibniz, *Die Philosophischen Schriften*. Ed. C. I. Gerhardt. 7 vols. Berlin: Weidmannsche Buchhandlung, 1875–90. Reprint, Hildesheim: Olms, 1960–1. Cited by volume and page.
Grua	G. W. Leibniz, *Textes inédits d'après les manuscrits de la Bibliothèque Provinciale de Hanovre*. Ed. G. Grua. 2 vols. Paris: PUF, 1948. Cited by page.
Guerrier	Woldemar Guerrier, *Leibniz in seinen Beziehungen zu Russland und Peter dem Grossen*. St Petersburg and Leipzig, 1873. Cited by page.
NE	G. W. Leibniz, *New Essays on Human Understanding*. Ed. and trans. Peter Remnant and Jonathan Bennett. Cambridge: Cambridge University Press, 1996.
PW	*The Political Writings of Leibniz*. Trans. and ed. with an introduction by Patrick Riley. Cambridge: Cambridge University Press, 1972.
WF	*Leibniz's New System and Associated Contemporary Texts*. Trans. and ed. R. S. Woolhouse and Richard Francks. Oxford: Clarendon Press, 1997.

Unless otherwise stated, translations are my own.

References

Chapter 1

The citation from the letter to Luise von Hohenzollern of 1705 is from J. G. H. Feder (ed.), *Commercium epistolicum*. Hanover, 1805, pp. 476–7.

Chapter 2

The citation from the letter to Newton of 10 December 1716 is from *The Correspondence of Isaac Newton*, 7 vols. Cambridge: Cambridge University Press, 1959–77 (vol. 6, pp. 376–7).

Chapter 4

The citation 'have a reality beyond our intelligence' is from Leibniz, *Notationes quaedam ad Aloysii Temmik Philosophiam* (c.1715–16), in Massimo Mugnai, *Leibniz' Theory of Relations*. Stuttgart: Steiner, 1992, p. 155.
The translations of the citations from A VI, 4, 17 and A VI, 4, 18 are by R. M. Adams in *Leibniz: Determinist, Theist, Idealist*. Oxford: Oxford University Press, 1994, pp. 179–80.

Chapters 8 and 9

The translation of the citation from GP II, 252 is by Adams in *Leibniz*, p. 265.

Chapter 8

The translation from Dutens, III, 499 is by Donald Rutherford, in 'Monads', *Oxford Handbook of Leibniz*, online version (DOI: <10.1093/oxfordhb/9780199744725.013.26>).

Further reading

Writings by Leibniz

Leibniz wrote in Latin, French, and German. Publication of the critical edition of his writings and correspondence is still ongoing and is expected to embrace some 120 large volumes. Known also as the 'Akademie Ausgabe', it is published by the Berlin Academy of Sciences in eight series under the general title of *Sämtliche Schriften und Briefe* (see 'Abbreviations' for full bibliographical details). Recent volumes are openly accessible online at <http://www.leibnizedition.de/>. This website also includes invaluable research tools for scholars working on Leibniz. A variety of other editions are still the reference point for important texts or correspondences not yet included in the Akademie Ausgabe. Some of the major editions are listed in the 'Abbreviations'. Among the English translations of selected writings and correspondences, the ongoing Yale Leibniz Series is a superb resource. The volumes contain both the original language text and an English translation on facing pages. They are edited by leading scholars and prefaced by substantial introductory essays providing marvellous discussions of the body of texts collected in the volumes.

Writings on Leibniz

The literature on Leibniz is vast. A comprehensive, continuously updated, and easily searchable database is available at <http://www. leibniz-bibliographie.de/>. On Leibniz's life, works, and intellectual development in its historical contexts, see Maria Rosa Antognazza, *Leibniz: An Intellectual Biography* (Cambridge: Cambridge

University Press, 2009). State-of-the-art essays covering the main aspects of Leibniz's thought (including philosophy, mathematics, science, technology, politics, and theology) are collected in the *Oxford Handbook of Leibniz* (Oxford: Oxford University Press, forthcoming). Individual essays are already available online at <http://www.oxfordhandbooks.com/>.

On Leibniz's logic, logical calculus, and theory of relations, the outstanding work of Massimo Mugnai, on which this book draws, is fundamental. In English, see especially his *Leibniz' Theory of Relations* (Stuttgart: Steiner, 1992); 'Leibniz's Ontology of Relations: A Last Word?', *Oxford Studies in Early Modern Philosophy* 6 (2012): 171–208; and his chapters in the *Oxford Handbook of Leibniz*. Leibniz's metaphysics and philosophical theology is magisterially discussed by R. M. Adams in *Leibniz: Determinist, Theist, Idealist* (Oxford: Oxford University Press, 1994), to which this book is also deeply indebted.

Another milestone monograph which shaped the treatment of theodicy and moral philosophy above is Donald Rutherford, *Leibniz and the Rational Order of Nature* (Cambridge: Cambridge University Press, 1995). On Leibniz's monadology, see Rutherford's illuminating essay on 'Monads' for the *Oxford Handbook of Leibniz*. On Leibniz's moral philosophy, see also Gregory Brown's insightful contributions: these include 'Disinterested Love: Understanding Leibniz's Reconciliation of Self- and Other-Regarding Motives', *British Journal of the History of Philosophy* 19 (2011): 265–303 and the chapter on 'Happiness and Justice' in the *Oxford Handbook of Leibniz*. Important collections of essays re-evaluating various facets of Leibniz's theodicy have recently appeared, including *New Essays on Leibniz's Theodicy*, edited by Larry M. Jorgensen and Samuel Newlands (Oxford: Oxford University Press, 2014), and *Lectures et interprétations des Essais de théodicée de G. W. Leibniz*, edited by Paul Rateau (Stuttgart: Steiner, 2011). On Leibniz's political philosophy and his ideal of justice as charity of the wise, see Patrick Riley's *Leibniz's Universal Jurisprudence: Justice as the Charity of the Wise* (Cambridge, Mass.: Harvard University Press, 1996).

Leibniz's theory of knowledge is discussed in excellent contributions by Martha Bolton (see e.g. her chapter in the *Oxford Handbook of Leibniz*). Leibniz's debate with Locke in the *New Essays* is the subject of an important monograph by Nicholas Jolley (*Leibniz and Locke: A Study of the New Essays on Human Understanding* (Oxford: Clarendon Press, 1984)). A number of volumes have been devoted to key philosophical correspondences, including Paul Lodge (ed.), *Leibniz and his Correspondents* (Cambridge: Cambridge University Press, 2004); R. C. Sleigh, *Leibniz and Arnauld: A Commentary on their Correspondence* (New Haven: Yale University Press, 1990); Ezio Vailati, *Leibniz and Clarke: A Study of their Correspondence* (Oxford: Oxford University Press, 1997); Brandon Look, *Leibniz and the 'Vinculum Substantiale'* (Stuttgart: Steiner, 1999). Highly recommended are also the extensive and ground-breaking introductions by Paul Lodge to his edition of the Leibniz–De Volder correspondence (New Haven: Yale University Press, 2013), and by Brandon Look and Donald Rutherford to their edition of the Leibniz–Des Bosses correspondence (New Haven: Yale University Press, 2007).

The debate on whether Leibniz was an idealist, a realist, or both, concerning bodies, and whether the development of his thought on this issue was broadly consistent, has generated a staggering amount of literature, discussing his metaphysics and philosophy of nature in great detail. Among those proposing an idealist reading of Leibniz's mature and/or final metaphysics, see especially the leading studies of R. M. Adams and Donald Rutherford.

The case for the presence in Leibniz of two different metaphysical models—one having as its fundamental entities quasi-Aristotelian corporeal substances, irreducible to monads; the other reducing to mind-like simple substances or monads—has been forcefully made by Daniel Garber in his brilliantly written and widely debated *Leibniz: Body, Substance, Monad* (Oxford: Oxford University Press, 2009). In *Leibniz and the Natural World* (Dordrecht: Springer 2005), Pauline Phemister challenges idealist interpretations. According to her proposal, the real metaphysical constituents of Leibniz's universe are not mind-like immaterial monads but indivisible, living, animal-like corporeal substances. Peter Loptson and R. T. W. Arthur argue for the robustness of Leibniz's commitment to the reality of body in 'Leibniz's Body Realism: Two Interpretations', *The Leibniz Review* 16 (2006): 1–42.

Catherine Wilson, *Leibniz's Metaphysics: A Historical and Comparative Study* (Princeton: Princeton University Press, 1989) mounts a sophisticated case against the view that Leibniz had a broadly coherent and unified philosophical system, identifying instead three incompatible metaphysical schemes held by Leibniz at the same time, although not in a fully conscious way. On the other hand, Christia Mercer's massive study on *Leibniz's Metaphysics: Its Origins and Development* (Cambridge: Cambridge University Press, 2001) stresses continuity, strongly arguing that the methodological and metaphysical commitments that Leibniz developed during his youth formed the bedrock of his mature philosophy. Finally, according to Glenn Hartz, *Leibniz's Final System* (New York: Routledge, 2007), Leibniz consciously endorsed both idealism and realism as 'theories of Reality' rather than 'absolute pronouncements about Reality'.

On the controversy over the discovery of the calculus and Leibniz's relationship with Newton see the seminal studies of Rupert Hall, *Philosophers at War: The Quarrel between Newton and Leibniz* (Cambridge: Cambridge University Press, 1980) and Niccolò Guicciardini, *Reading the Principia: The Debate on Newton's Mathematical Methods for Natural Philosophy from 1687 to 1736* (Cambridge: Cambridge University Press, 1999), on which the foregoing discussion draws, as well as Domenico Bertoloni Meli, *Equivalence and Priority: Newton versus Leibniz* (Oxford: Clarendon Press, 1993).

Publisher's acknowledgements

We are grateful for permission to include the following copyright material in this book:

Extracts taken from *The Political Writings of Leibniz*. Translated and edited with an introduction by Patrick Riley (Cambridge University Press, 1972). By permission of Cambridge University Press.

Extracts taken from G. W. Leibniz, *New Essays on Human Understanding*. Translated and edited by Peter Remnant and Jonathan Bennett (Cambridge University Press, 1981). By permission of Cambridge University Press.

Extracts taken from R. M. Adams, *Leibniz: Determinist, Theist, Idealist* (Oxford University Press, 1994). By permission of Oxford University Press.

Extracts taken from *Leibniz's New System and Associated Contemporary Texts*. Translated and edited by R. S. Woolhouse and Richard Francks (Clarendon Press, 1997). By permission of Oxford University Press.

The publisher and author have made every effort to trace and contact all copyright holders before publication. If notified, the publisher will be pleased to rectify any errors or omissions at the earliest opportunity.

Index

Leibniz

Index

Leibniz

SOCIAL MEDIA
Very Short Introduction

Join our community

www.oup.com/vsi

- Join us online at the official Very Short Introductions
  Facebook page.
- Access the thoughts and musings of our authors with our
  online **blog**.
- Sign up for our monthly **e-newsletter** to receive information
  on all new titles publishing that month.
- Browse the full range of Very Short Introductions online.
- Read **extracts** from the Introductions for free.
- If you are a teacher or lecturer you can order inspection
  copies quickly and simply via our website.

ONLINE CATALOGUE
A Very Short Introduction

Our online catalogue is designed to make it easy to find your ideal Very Short Introduction. View the entire collection by subject area, watch author videos, read sample chapters, and download reading guides.

http://global.oup.com/uk/academic/general/vsi_list/